SpringerBriefs in Earth System Sciences

Series Editors

Gerrit Lohmann, Universität Bremen, Bremen, Germany

Justus Notholt, Institute of Environmental Physics, University of Bremen, Bremen, Germany

Jorge Rabassa, Labaratorio de Geomorfología y Cuaternar, CADIC-CONICET, Ushuaia, Argentina

Vikram Unnithan, Department of Earth and Space Sciences, Jacobs University Bremen, Bremen, Germany

SpringerBriefs in Earth System Sciences present concise summaries of cutting-edge research and practical applications. The series focuses on interdisciplinary research linking the lithosphere, atmosphere, biosphere, cryosphere, and hydrosphere building the system earth. It publishes peer-reviewed monographs under the editorial supervision of an international advisory board with the aim to publish 8 to 12 weeks after acceptance. Featuring compact volumes of 50 to 125 pages (approx. 20,000–70,000 words), the series covers a range of content from professional to academic such as:

- A timely reports of state-of-the art analytical techniques
- bridges between new research results
- snapshots of hot and/or emerging topics
- literature reviews
- in-depth case studies.

Briefs are published as part of Springer's eBook collection, with millions of users worldwide. In addition, Briefs are available for individual print and electronic purchase. Briefs are characterized by fast, global electronic dissemination, standard publishing contracts, easy-to-use manuscript preparation and formatting guidelines, and expedited production schedules.

Both solicited and unsolicited manuscripts are considered for publication in this series.

Theophilus Clavell Davies
Editor

Recent Advances in Medical Geology Research in Africa: A Decadal View

 Springer

Editor
Theophilus Clavell Davies
Faculty of Applied and Health Sciences
Mangosuthu University of Technology
Umlazi, KwaZulu Natal Province, Republic
of South Africa

ISSN 2191-589X   ISSN 2191-5903 (electronic)
SpringerBriefs in Earth System Sciences
ISBN 978-3-032-13753-1   ISBN 978-3-032-13754-8 (eBook)
https://doi.org/10.1007/978-3-032-13754-8

Preface

The first edition of the book: *Recent Advances in Research on Medical Geology of Africa: A Decadal View* puts together a series of selected peer-reviewed papers presented at an "Organised Session" of The Joint Conference of ISEH, ICEPH and ISEG on Environment and Health", held in Galway, Republic of Ireland from 11 to 18 August 2024. The objective of the Organised Session was to provide a platform to review the current status of Medical Geology research in Africa, document the successes, identify the knowledge gaps, address the pitfalls and challenges; and take a look towards the future. It is against this backdrop that the design of further research on the subject would be predicated which also is in concordance with the underlining theme, geared towards the maintenance of the Continent's public health integrity.

The first and most important question potential readers should have about this book is: How is it different from other Medical Geology books on the market? Well, the book presents some of the latest (last 10 years) developments on key topical issues in the field of Medical Geology of Africa in three ways: integration and analyses of some of the latest research; identification of knowledge gaps, and proffering credible ways of addressing the challenges through adoption of the recommended interventions and application of the new technologies that have been developed. Although these advancements are captured by many relevant journals, this book highlights the advances in a more systematic way, incorporating eight chapters contributed by researchers of the Mangosuthu University of Technology (MUT).

Most environmental diseases of concern to the Medical Geologist are geochemical diseases brought about by various positive and negative systemic interactions involving trace elements/metals/metalloids/isotopes after different types of exposure episode. Therefore, success in researching geochemical diseases can be said to be contingent on the accuracy of determination of geomedically important chemical elements [sometimes down to parts per billion (ppb) level]. The recent improvements in analytical capacity of African Geochemical Laboratories therefore have the potential to accelerate research progress on geochemical diseases in Africa.

An updated bibliography is assembled for each chapter, to ease the burden of researchers seeking the most up-to-date information on the topics covered; and in setting the groundwork for their further research into the exposed knowledge gaps.

The bibliographic compilation targeted relevant peer reviewed articles published during the last decade or so.

Chapter 3, "Criticality of the Optimal Intake Level of Nutritional and Potentially Toxic Elements in the African Food Chain" is particularly noteworthy, for it highlights the tremendous effort being expended recently on understanding the processes and mechanisms of uptake of nutritional and potentially harmful elements (PTEs) from the soil through food crops, and their role in mediating metabolic processes and influencing immune system functions. As the link between geochemical cofactors and certain environmental diseases becomes clearer and more firmly established, we will continue to see a rise in research activities in this field.

The book is of relevance to researchers and graduate students of African Medical Geology and Geoenvironmental Medicine who wish to understand the challenges associated with each topic; as well as other members of the scientific community wishing to enhance their knowledge of these emerging fields. Readers are encouraged to contact the authors to establish collaboration opportunities.

Chapter 7, "A Code for Medical Geology Fieldwork in Africa: Guidelines on Health and Safety Issues in Mapping Disease Distribution and Their Geoenvironmental Correlates" has been prepared for all stakeholders involved with geological fieldwork in Africa, with the focus predominantly, but not exclusively, at postgraduate students of Medical Geology carrying out their field projects, and their fieldwork leaders (FWLs). Its aim is to encourage geomedical science and related departments having substantial fieldwork components in their curriculum, to adopt a risk-based approach to the management of health and safety issues arising out of fieldwork and set out reasonably practicable actions.

Chapter 8, "Research Gaps in Knowledge and Future Perspectives on Medical Geology of Africa", the conclusion chapter briefly summarises the successes, lists the challenges that remain and takes a circumspect look at future perspectives.

As research and study of Medical Geology in Africa continues to draw the fascination of hundreds of graduate students and researchers in allied disciplines, it is hoped that the advent of this book will contribute towards keeping this momentum going!

Umlazi, Republic of South Africa   Theophilus Clavell Davies

Acknowledgements

The chapters in this book are based on selected peer-reviewed papers presented at an Organised Session on: "Recent Advances in Research on Medical Geology of Africa: A Decadal View" of "The Joint Conference of ISEH, ICEPH and ISEG on Environment and Health" held in Galway, Republic of Ireland from 11 to 18 August 2024. I wish to thank the Conference Organising Committee for their superb organisational skills and great circumspection in completing the papers for this Session.

The Research, Innovation and Engagements Directorate of Mangosuthu University of Technology (MUT) of South Africa supported various stages of the book project, including financial and logistical arrangements for travel and presentation of the corresponding conference papers in Galway for all MUT researchers. Other financial resources covering data acquisition and analytical work, as well as writing and publication, were sought from research grants from the "South Africa Highly Disadvantaged Institutions (HDI) Subvention" and the South Africa "National Research Foundation (NRF) Incentive Funding for Rated Researchers" (IFRR).

Any copyrighted material on these pages is used under "fair use" for the purpose of study and review. A thorough effort was made to clear all necessary reprint permissions. All illustrations and citations in the text are credited to their appropriate and respective sources, and a general expression of my indebtedness and thanks is given to the many institutions and authors who have permitted their images and photographs to be used in this book. Any required acknowledgement omitted is unintentional.

I thank the (anonymous) peer reviewers who read the manuscript for raising important queries, and for providing many insightful comments and suggestions. The careful attention paid towards addressing these concerns has resulted in a much improved manuscript quality.

Finally, I wish to express my sincere thanks to the staff of my publishers, Springer Nature Switzerland AG for their enthusiastic and helpful co-operation throughout the production of this book; and, in particular, Ashok Arumairaj, Annett Büttner, Coral Zhou, A. P. Umamagesh and other editorial staff members, for their professionalism, diligence and patience.

Contents

Chapter 1
Introduction—Recent Advances in Research on Medical Geology of Africa: A Decadal View

T. C. Davies

Medical Geology studies the interrelationship between geological materials and processes and the health of man and animals, with the goal of deciphering credible ways of combatting these diseases. The chapters in this book synthesise the current status of some critical aspects of the Medical Geology research around the Continent during the last decade, documents the successes, identifies the knowledge gaps, addresses the pitfalls and examines the current challenges and future perspectives. It is against this backdrop that the design of further research on the subject would be predicated.

The surface geochemistry of Africa is unique and extremely complex. Separation, remobilisation and redistribution of elements by surficial geogenic processes (such as intense tropical weathering, leaching and lateritisation) have created large areas exhibiting deficiency in some key nutritional elements, and excessive amounts in others. Similarly, large amounts of potentially harmful elements (PHEs) are dispersed widely in some arable soils, especially those associated with mineralised terrains.

The upsurge in Medical Geology research in Africa during the last decade has been driven chiefly by awareness of the impact of this heterogeneity in surface geochemistry of the Region on the food quality of agricultural crops. There is also the increased realisation of the criticality of the optimal intake levels of nutritional elements and the importance placed on the abeyance of PHEs from the food chain for maintaining health.

Environmental pollution (the subject of Chap. 2) is perhaps one of the largest causes of disease and death in Africa today (2025). Yet this kind of pollution and the harmful effects on the people and ecosystems have not been given due attention by the Region's governments; and a truly regional policy response still goes abegging.

T. C. Davies (✉)
Faculty of Applied and Health Sciences, Mangosuthu University of Technology, Umlazi, KwaZulu Natal Province, Republic of South Africa
e-mail: daviestheophilus2025@yahoo.com

T. C. Davies (ed.), *Recent Advances in Medical Geology Research in Africa: A Decadal View*, SpringerBriefs in Earth System Sciences, https://doi.org/10.1007/978-3-032-13754-8_1

Chapter 2 also summarises the decadal results of researchers on the role of the physicochemical properties of the different environmental media (soil, water and air) that influence disease distribution around the Continent. This chapter lists a series of measures to combat environmental pollution on the continent and reviews the relatively new technologies developed to combat the problem.

Sufficiency of intake levels of nutritional elements does not guarantee availability of the nutrient in the metabolic pathways. Chapter 3 examines the relationship between the geochemical fluxes of both nutritional and PTEs in the groundwater-soil-food crop continuum and the nutritional quality of the African diet, and hence the correct balance of micronutrients in metabolic processes. The main highlight in this chapter, then, is on how the criticality of the intake levels of nutritional elements, in particular, the amount that reaches the metabolic centres of activity, which in turn, is determined by biophysicochemical parameters such as bioavailability, bioaccessibility and dose–response relationships. It is this amount of the element (nutritional or PTE) that eventually gets into the metabolic centres that determines whether nutritional benefit is realised or disease is produced. Such effects can be expressed in a *dose–response curve*. Examples are given of public health conditions that may reflect nutrient poor environments in Africa including iodine deficiency disorders (IDD), fluoride (F^-) deficiency/excess symptoms, maladies due to geophagic practices, the selenium deficiency/HIV-AIDS postulation and maladies associated with iron deficiency or excess.

In South Africa, as in other mineral-rich African countries, artisanal and small-scale mining (ASM) is crucial to the livelihoods of hundreds of inhabitants including a significant proportion of women, particularly in the gold mining areas of the country. However, artisanal and small-scale gold mining (ASGM) is largely informal, and alternative economic gains can lead to health degradation, and can heave untold damage on vital industries and biodiversity. Substantial quantities of mercury (Hg) are still being used for the amalgamation of gold, despite the continued introduction of more efficient mining techniques and new Government legislations. Using South Africa in an illustrative case description, the focus of Chap. 4 is on how the application of geochemical techniques in source identification of Hg, quantification of emissions and decipherment of transport mechanisms can provide much needed data for the formulation of measures aimed at minimising the spread of Hg pollution; and so, preserve the environmental health of the miners and associated communities.

Emission of geogenic dust can contribute to air pollution and has been linked to increased mortality rates. For example, the Sahara Dust Storms (SDS) and rising warm air can lift dust thousands of kilometres or so above the African deserts and then out across the Atlantic, reaching as far as the Caribbean. The SDS can significantly impact health, particularly for those with pre-existing respiratory conditions like asthma or chronic obstructive pulmonary disease (COPD).

Chapter 5 examines the latest research (largely post-2015) on geogenic dust exposure and toxicity in Africa in order to obtain a greater understanding of the mechanisms and pathogeneses of dust-induced maladies; and so facilitate the design of effective dust control measures and clinical procedures for halting disease progression. Dust also plays a role in climate change, with different particle sizes having varying effects on atmospheric heating and cooling. The final part of Chap. 5 therefore looks at the intersection of climate change and dust pollution.

Based on published data and epidemiological records, Chap. 6 illustrates how climate induced perturbations in frequency and intensity of geological processes in Africa's coastal zone, have impacted the spectrum of health risks; with some attention paid to the effects on future regional food security. The chapter also presents several key recommendations in addressing food insecurity stemming from climate change, focussing on enhancing climate resilience in agriculture, promoting diversified livelihoods and strengthening disaster preparedness and response mechanisms.

Geoscientific and geoecological fieldwork undertaken by students, just like other categories of fieldworkers, often involves significant threats to health and safety. Field education and mapping programmes can lead a group of researchers to some of the wildest and most inhospitable parts of Africa, and every field situation is different. Instructors and assistants should take the time to analyse specific regional risks and other unique issues. For example, many fieldwork leaders on trips in Africa have only rudimentary knowledge of how to treat snake bites. It is impossible to know everything that might happen to researchers or their leaders in the field, but they should expect (and prepare for) the unexpected. Chapter 7 reiterates the need for careful planning and preparation which can mitigate situations that could affect the health and safety of our Medical Geology students and allied researchers to ensure the overall success of field programmes.

In Chap. 8, the conclusion Chapter, a summary of the key findings is presented, together with a synthesis of the work into a broader context. A circumspective look is taken of the prospects for future development of Medical Geology in Africa.

Chapter 2
Current Status of Research on Geogenic Pollution and Disease in Africa

T. P. Makhathini and T. C. Davies

Abstract African communities face significant health challenges related to environmental pollution through various forms such as soil degradation from pollutants, water contamination, airborne toxins, improper waste disposal practices and global warming phenomena. Processes causing environmental contamination encompass the geochemical circulation of naturally occurring elements as well as those introduced by human activities; they may result in numerous health issues and socioeconomic problems. The aim of this Chapter revolves around the synthesis and evaluation of recent studies linking *geogenic* pollutants in soil, water and air directly to various environmental health issues affecting Africans. Although numerous research efforts have been dedicated to addressing disease prevention through methods related to soil, water and air pollution across Sub-Saharan Africa, significant gaps in knowledge remain regarding effective pollution control strategies as well as optimal approaches for combatting resultant illnesses. Addressing these knowledge deficiencies demands an extensive comprehension of how different pollutants interact and their routes into ecosystems. This comprehensive insight will broaden our ability to diagnose numerous African environmental health issues effectively and guide us towards sound solutions.

Keywords Geogenic pollution · Environmental health · Diseases · Research gaps · Addressing the issues · Africa

T. P. Makhathini
Faculty of Engineering, Mangosuthu University of Technology, Umlazi, KwaZulu Natal Province, Republic of South Africa

T. C. Davies (✉)
Faculty of Applied and Health Sciences, Mangosuthu University of Technology, Umlazi, KwaZulu Natal Province, Republic of South Africa
e-mail: daviestheophilus2025@yahoo.com

Introduction

African environmental pollution problems come from both natural sources and human actions. In this chapter, the term "geogenic sources" refers to "naturally occurring" as a result of geological processes such as the release of harmful substances into the environment through seismic activity or volcanic action. Other examples of geogenic sources of pollution include weathering, the emission of windblown dust, emanations from mineralised terrains and the natural presence of toxic metals in the land. These naturally occurring pollutants can cause various health conditions and maladies for people and the environment.

Human activities (anthropogenic) also contribute to pollution. For example, burning fossil fuels and the waste from landfills can release harmful substances into the environment. Both geogenic and anthropogenic pollution can greatly affect human health, animals and the ecosystems around us.

The chapter also reviews recent research that deals with these problems and points out gaps in our understanding of how pollution starts, how it affects people and how it moves through the environment. It also gives directions for future research in this area.

In Lucero-Prisno et al. (2023), and others listed 10 important areas for public health research, with environmental pollution being a major one. They stressed the need for good strategies and new, proactive health practices to address this issue.

The main sources of geogenic pollution in Africa include toxic elements like arsenic, cadmium, fluorine, lead and mercury. These elements are often released through mining activities and natural events such as earthquakes, volcanoes and geothermal processes. A lot of these pollutants also come from human activities like farming and waste disposal. Radon gas, which comes from uranium and gold mining is a growing concern in countries like Namibia, Niger and South Africa (Table 2.1).

In Africa, reports on mortality and morbidity due to environmental pollution are legion and dire. For example, the United Nations Children's Fund (UNICEF) estimates the number of children under five who will die from air pollution-related causes in South Africa during 2021 (HEI/IHME 2024) at more than 3,365.

Nature and Sources of Soil Pollution in Africa

Soil pollution in Africa originates from both geological and anthropogenic sources. Toxic metals, radionuclides, asbestos and many other contaminants may be present in soils due to geological and pedological processes with or without anthropogenic impact (FAO and UNEP 2021; Xin et al. 2022). Purely geological processes occurring in many areas of Africa (e.g., volcanic eruptions, earthquakes, weathering and erosion, floods and tsunamis) produce soil pollution (Tindwa and Singh 2023).

Table 2.1 Examples of major sources of geogenic pollution in some regions of Africa

Type of pollutant	Source of pollution	Reference
Arsenic and fluoride contamination	Many African nations have naturally occurring groundwater containing arsenic and fluoride, especially in areas with volcanic rocks, mineralised zones, and geothermal activity. Because of the interplay between water and volcanic rock formations, the Rift Valley regions of East Africa—most notably Ethiopia, Kenya and Tanzania—are infamous for having above normal fluoride concentrations in their groundwater systems	Demelash et al. (2019), Nordstrom (2022), Kapwata et al. (2023), Ashong et al. (2024), Usman et al. (2024)
Toxic metal pollution	Significant environmental contamination with toxic elements is caused by mineral exploitation, mining and ore processing in mineral-rich countries such as Ghana, Nigeria, South Africa and Zambia. For instance, mining activities in the Zambian Copperbelt have been linked to lead and copper poisoning of water and soil, with grave health repercussions for people living in the vicinity of the mining centres. Massive mercury pollution has resulted from artisanal gold mining in some countries of West and southern Africa	King et al. (2024), Musiige et al. (2024), Ripanda et al. (2025)
Radon	High concentrations of radon, a radioactive gas emitted during uranium decay, are found in uranium-rich soils in farmlands adjoining major uranium and gold mining centres of Sub-Saharan Africa. Significant knowledge gaps still exist in our knowledge of radon distribution in rocks, soil and natural waters of Africa and its health effects	Mbuende (2022), Mphaga et al. (2024), Rathebe et al. (2025)

Table construction done by the authors

According to Fayiga et al. (2018), other sources of soil pollution in Africa that may be classed as anthropogenic include agricultural activities, roadside emissions, auto-mechanic workshops, garbage dumps and e-waste. These sources of soil pollution are also important in terms of disease causation. An example of how overuse of soil for agriculture, along with erosion caused by rain, rivers and winds, can lead to soil becoming infertile is presented in the plains of the Nile River which runs through the northern and eastern plains of Africa and the Orange River in southern Africa.

Soil Pollution Through Mining and Processing of Ore

Mining and ore processing are important economic activities in many African coun-tries but are also major sources of environmental pollution. The large quantities of waste generated through these processes make the Continent liable to wide-scale pollution of the soil, water and air environment. Raimi et al. (2022) place 'mining' at the top of their listing of contaminant sources of soil pollution.

The large portions of toxic metals and other poisonous elements delivered in the soil surroundings can persist for lengthy periods, long after mining or ore processing activities have ceased (Ogundele et al. 2017). Mining wastes are stocked up in tailings comprised of fine particles containing huge amounts of toxic metals.

Geogenic Sources of Water Pollution in Africa

Water pollution in Africa is worsening, with mining being regarded as one of the predominant causes of this disaster. The immoderate use of water in ore processing operations, the careless discharge of mine effluents and seepage from tailings and waste rock impoundments are the principle reasons for pollution of freshwater resources in Africa.

In Ighalo et al. (2021), systematically analysed the published literature and typified the nature and local distribution of pollution sources in Africa within the realm of surface water, groundwater and rainwater quality. This author used Nigeria in a case description. This is a country that has for long been confronted with a massive hassle of water pollution wrought through petroleum resources development.

Some other extensive groundwater pollution problem in Africa comes from leachate discharge from landfills and dumpsites, with organics, toxic metals and salts constituting the greater percentage of pollutants from this source (see, e.g., Bruce and Limin 2021; Igboama et al. 2022).

Water Pollution Through Mining

In recent years, there have been some improvements in mining practices in Africa (see, for example, GAN 2024); but there are still substantial environmental risks. The effect of past mining activities on water systems is still a major concern. Because mining has become more mechanised, more rock and ore are being moved than ever before. This has led to four main types of water pollution (SDWF Updated 2023).

(1)   Acid Mine Drainage

The term "Acid Rock Drainage" (ARD) refers to a natural process. When sulphides in rocks are exposed to air and water, they create sulphuric acid. "Acid Mine Drainage" (AMD) is similar, but much worse. When the acidity becomes strong enough, a type of bacteria called *Thiobacillus ferroxidans* becomes active. These bacteria speed up the reaction, making more acid and causing more metals to be released from the waste. As long as the rocks are exposed to air and water, the acid continues to leach out of the rock. This can go on for hundreds or even thousands of years. The acid is carried away by rain or surface water into nearby rivers and groundwater. This can ruin water quality and harm aquatic life, making water unsafe to use.

The chemistry behind AMD is complex, which makes it hard to find effective remedies for the phenomenon. Many chemical processes are involved; but the primary process involves the oxidation of pyrite.

The following equation can be used to describe AMD formation:

$$2FeS_2(s) + 7O_2(g) + 2H_2O(I) = 2Fe^2 + (aq) + 4SO_4^{2-}(aq) + 4H^+(aq)$$

AMD can cover large areas, especially in tailings storage facilities (TSF), because wind and water can move the materials around (see: Laker 2023).

AMD is of great concern in South Africa, especially in the Bushveld Complex, where many metal sulphide mines are located. This has caused a lot of disquiet among environmental groups.

(2)   Toxic Metal and Metalloid Contamination and Leaching

Toxic metals and metalloids, such as cobalt, copper, cadmium or arsenic, can pollute water when they come into contact with it. These substances are carried away by water that flows over exposed rock. The acid in streams draining tailings dumps can hold elevated levels of sulphate and heavy metals like lead, mercury, zinc and platinum. Arsenic is also commonly found in these areas (Irunde et al. 2022).

(3)   Processing Chemicals Pollution

When chemicals used in mining, such as cyanide or sulphuric acid, leak or spill into nearby waterways, they cause chemical pollution. These substances are used to extract minerals from the ore.

(4) Erosion and Sedimentation

Diverse human activities such as construction work and maintenance of roads, open pits and waste impoundments during mineral development, can result in the disturbance of soil and rock piles. If proper measures are not taken, erosion can carry large amounts of soil and rock into nearby waterways. Too much sediment can block rivers, damage plant life, and harm animals and their habitats.

Other Sources of Water Pollution

Other sources of water pollution in Africa that are important include waste from cities and industries, landfills, sewers, roads and car parks. Waste that is not treated properly gets washed into drains and then flows into nearby rivers and waterways, causing widespread pollution and health threats to humans and animals. The most common natural disaster in Africa is flooding.

Flooding occurs when water covers land that is normally dry. This usually happens because of heavy rain, fast melting of snow, a dam breaking or storm surges from tropical storms or tsunamis in coastal areas. Flooding often leads to a wide range of water-related diseases (see Section on: "Environmental health impacts of water pollution in Africa", this Chapter).

Hydraulic fracturing, also known as *fracking*, is a method used to get oil and gas from shale rock. It is a major source of water pollution in countries where it is being used, like South Africa. During this process, a lot of wastewaters is produced and stored in tanks. Sometimes, this water leaks into the ground, polluting the groundwater and making it unsafe for drinking.

The Nature and Sources of Air Pollutants

Air pollutants can be primary or secondary. They include very small particles called $PM_{2.5}$, which are 2.5 μm or smaller in size. These particles come from both natural and human-made sources. Natural sources include windblown dust from roads and drysoil, sea spray and wildfires. Human-made sources include burning of fossil fuels and biofuels, transportation, industrial activities like small-scale mining, farming and burning of waste. When soil is left exposed, especially in dry areas, the wind picks it up, creating dust and fine particles that become part of the air as aerosols. These particles can range from very fine to coarser materials such as silt.

Environmental Health Impacts of Soil Pollution in Africa

In 2023, Tindwa and Singh noted that the impact of soil pollution on human health in the Sub-Saharan Africa region is not adequately reported; but the effects are thought to be diverse, both indirectly through the consumption of contaminated food and drinking water, and directly through contact with contaminated soil (FAO and UNEP 2021). *Geophagy* or *geophagia*, the intentional or sometimes unintentional consumption of soil or related earth materials is a common practice in Africa. It can lead to a variety of diseases (see: Davies 2023). Soil pollution also affects soil fertility, which threatens food security.

Soil pollution can cause diseases whose effects can range from non-fatal to life-altering. In 2015, Oliver and Gregory identified six soil-related human health risks, three of which are related to soil pollution, namely: risks from elemental contamination (such as arsenic, cadmium and lead); organic chemical contamination [e.g., polychlorinated biphenyls (PCBs), polycyclic aromatic hydrocarbons (PAHs), persistent organic pollutants (POPs)]; and pharmaceutical contamination (e.g., oestrogen, antibiotics). Three other risks according to these authors (Oliver and Gregory 2015) result from exposure to soil pathogens such as anthrax and prion, micronutrient deficiencies and undernutrition due to poor soils.

Environmental Health Impacts of Water Pollution in Africa

Many communities in Africa lack access to safe, clean water for drinking, cooking and sanitation. Many recorded illnesses, diseases and deaths have been attributed to water shortages and pollution of water resources from both geologic and anthropogenic sources.

Poor sanitation practices have resulted in an abundance of waterborne tropical diseases, such as typhoid fever, cholera, dysentery and diarrhoeal diseases, occurring at different times of the year in different parts of the Continent (see: Lin et al. 2022). Plague, typhus and trachoma (an eye infection that can result in blindness) are also common conditions related to water pollution.

As population continues to grow and contributors such as urbanisation impact water bodies across the Continent, water scarcity and pollution are becoming worse. Geologic sources of water pollution, including mining-related processes and activities (for example, AMD), can deplete surface and groundwater supplies. Overexploitation of groundwater can damage or destroy streamside habitats many kilometres away from the actual mine site.

Air Pollution and Disease in Africa

Fuller et al. (2022) placed air pollution at the top of the list of environmental causes of disease and premature death globally, with Africa recording some of the worst cases of air pollution and some of the most severe health consequences compared to the rest of the world (Fig. 2.1). In 2019, air pollution was found to be the second leading risk factor for death across Africa after malnutrition (HEI 2022); and caused 1·1 million deaths, with the poor, the elderly and young children being the groups most at risk (see: Fischer et al. 2021; HEI 2022).

The range of diseases caused by regular inhalation of fine air pollution particles or aerosols ($PM_{2.5}$) is wide. They include ischemic heart disease, chronic obstructive pulmonary disease (COPD), stroke, lung cancer, type 2 diabetes and neonatal disorders (Wang et al. 2022; GBD 2019 Chronic Respiratory Disease Collaborators 2023). Several recent studies have proven that air pollution significantly contributes to poor brain health (for example, neurodegenerative diseases in adulthood and neurological effects in childhood and adolescence) as well as reduced life expectancy (i.e., the number of years a person can possibly expect to live) (HEI 2022).

Most roads in rural settings in Africa are unpaved, non-tarmac or constructed using lateritic soil; therefore, dust mixed with poisonous metallic and metalloid particulates (such as lead and arsenic) in addition to various pathogens, are freely inhaled causing bronchial asthma and various respiratory problems.

According to Fischer et al. (2021), the loss in financial output in 2019 due to morbidity and mortality associated with air pollution in some African countries was $3·02 billion (1·16% of GDP) in Ethiopia; $1·63 billion in Ghana (0·95% of GDP); and $349 million (1·19% of GDP) in Rwanda].

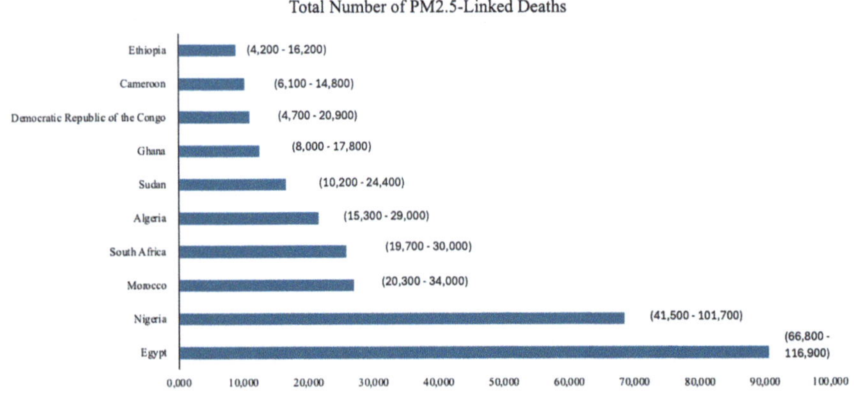

Fig. 2.1 Top 10 countries with the most $PM_{2.5}$-related deaths across Africa in 2019 (Image created by authors)

Health Effects of Geogenic Emissions and Emanations

Health Effects of Silica Dust

The term "*pneumoconiosis*" refers to a form of lung disorder caused by the inhalation of inorganic dust; and is characterised by the build-up of dust within the lungs and the reactions of the tissues to its presence. Silicosis is a long-term, occupational lung disorder that results from the inhalation of large amounts of dust containing tiny crystalline silica produced when crushing, drilling, blasting or grinding silica-containing rock. Although data on silicosis in the mining sector in southern Africa are limited, some current studies indicate that it is a major occupational health issue (e.g., Mojakwana and Ivuso 2023). Yi et al. (2023) have analysed the current disease burden of silicosis with the aim of gathering information needed to predict future growth trends of silicosis in the world, including southern Africa, by extracting records from the Global Burden of Disease (GBD) database (GBD 2019 Chronic Respiratory Diseases Collaborator 2023).

When the dust can be identified from coal handling and processing, the type of silicosis contracted is known as "*coal workers' pneumoconiosis*" (CWP), also known as *black lung disease* (see also under: "Health effects of silica dust", Chap. 5, this Volume). It is a pulmonary condition caused by prolonged exposure to a coal mining environment. This not only leads to high morbidity and mortality among miners but also imposes a heavy financial burden and diminishes the labour power within the society. The incidence of CWP as the major occupational lung disease in African coal mining areas is of great concern.

This is particularly so in South Africa, where postmortem studies have shown that prevalence rates among miners in the sixties were as high as 26% (Singh and Irusen 2025), although officially reported numbers may be lower as a result of reporting biases.

Taking into account the emerging developments in the international coal enterprise, the impacts of CWP are expected to evolve into new patterns and versions. Wang et al. (2024a) have provided a thorough assessment of the current panorama, fashion dynamics and future projections of the global CWP burden from 1990 to 2019.

In African coal mining environments, the association of CWP with tuberculosis (TB) is still a major health concern (Chakraborty et al. 2025; Singh and Irusen 2025). *However, despite the recognised frequent association of silicosis among miners with an increased risk of tuberculosis, the relationship between coal dust exposure and tuberculosis is still not fully understood* (Chakraborty et al. 2025).

Health Effects of Sahara Dust Particles

The health effects of geogenic emissions and emanations have been extensively studied in recent times. Impacts include air quality degradation by Sahara Dust (SD) (Kotsifakis et al. 2019; Sakhamuri and Cummings 2019; Harr et al. 2024), *although harmful organic pathways have not yet beenfully identified* (Bradek et al. 2024). Health hazards to humans can be acute, especially for people with lung conditions (see: "Health effects of Sahara dust plumes" and Chap. 5, this Volume). These issues include respiratory and cardiovascular diseases including life-threatening events in extreme cases (see, for example, Münzel et al. 2023; Pauri et al. 2024). Bradek et al. (2024) have added to this list the lung disease-related cytokines granulocyte–macrophage colony-stimulating factor (GM-CSF) and granulocyte colony-stimulating factor (G-CSF), both of which are important cytokines that participate in the development and progression of lung diseases, including bronchial asthma, chronic obstructive pulmonary disease (COPD) and pulmonary fibrosis.

Volcanic Gases and Health

Gases released from volcanoes may include sulphur dioxide (SO_2), hydrogen sulphide (H_2S), hydrogen chloride (HCl), hydrogen fluoride (HF), carbon dioxide (CO_2) and steam. For example, multigas measurements performed in February 2020 by Boudoir et al. (2022) at the volcanic gas plumes of Nyiragongo and Nyamulagira (Democratic Republic of the Congo), respectively, found average CO_2/SO_2 ratios of 22 ± 8.4 and 15.5 ± 5.7 and average H_2O/CO_2 ratios of 1.9 ± 0.2 and 12.4 ± 6.3.

Adverse health effects of exposure to volcanic gases vary from mild to severe; and sometimes it can also be fatal. Common effects of these gases, especially SO_2, H_2S, HCl and HF, are irritation of the respiratory tract, eyes and skin; chest tightness, shortness of breath and headache; and exacerbation of asthma. Sulphur dioxide is the biggest respiratory hazard, especially for people with asthma (UNDRR 2024). High concentrations of fluoride (from HF) cause damage to teeth and bones, especially in populations around the African Rift Valley (see, e.g., Marwa et al. 2018; Shaji et al. 2024); It is especially dangerous for grazing animals. All listed gas species, as well as CO and CO_2 can cause death in high concentrations. Inhalation of natural nanoparticles released from volcanic activity can also pose a threat to human health (see, e.g., Ermolin et al. 2021; Sonwani et al. 2021; Schiavo et al. 2023).

Health Effects of Fossil Fuel Combustion

Burning fossil fuels emits sulphate particles and SO_2 (gas), which can reflect sunlight and cool the atmosphere. In Kirk (2023), noted that, unlike volcanic aerosols, air pollution aerosols are continuously generated, but do not travel very high in the atmosphere. Inhalation of small particles emitted during the burning of fossil fuels causes asthma, respiratory infections, lung cancer and heart disease (Lelieveld et al. 2023). In Africa, the health impacts of fossil fuel consumption are severe. For example, in Gbenga Wilfred and Ohonga (2024) reported that, all things being equal, South Africa is projected to see a 0.418% increase in under-five mortality per 1000 persons per year for every 1% increase in fuel energy consumption.

Health Effects of Other Mineral Dusts

There are other mineral dusts that can cause undesirable pulmonary reactions but are not firmly established human carcinogens. These include kaolinite (*kaolinosis*), talc (*talcosis*) and iron oxide (*siderosis*) (see: Nemery 2022; Borm 2023). *The aetiology of diseases resulting from the inhalation of such minerals has been difficult to define, primarily because epidemiological investigations are complex, as they involve co-exposure to other minerals such as quartz and* asbestos (Huang et al. 2006; Fubini and Feneglio 2007; Filippova et al. 2022).

Asbestos Dust

The common designation *"asbestos"* refers to a group of six unique natural fibrous silicate minerals—actinolite, amosite, anthophyllite, chrysotile, crocidolite and tremolite. Occupational asbestos exposure in the general population has led to adverse health effects such as pulmonary fibrosis and mesothelioma; as well as restrictions on its use in countries such as South Africa that came too late to reduce exposure. Chapter 5 of this Volume discusses the mining of asbestos, restrictions on its use and the serious health effects of inhalation of asbestos dust.

The Current State of Air Pollution Research in Africa—A Paradox

Some of the highest levels of fine particles in the world have been recorded in cities of Sub-Saharan Africa (SSA) and other developing regions (Katoto et al. 2019; Fuller and Amegah 2022). *Yet, air pollution in Africa is a vastly under-researched topic.* This is partly due to the lack of data on the magnitude of air pollution concentrations and associated health effects in individual countries. According to Agbo et al. (2021), data available through 2020 showed that vehicle traffic, industry, forest fires and biomass burning are significant sources of particulate matter, carbon monoxide, nitrogen dioxide, sulphur dioxide and volatile organic carbon on the Continent.

Despite global call for urgent action against air pollution, many African countries still lack, as of the beginning of the current decade, functional air quality monitoring stations and data from them, making air quality management difficult (Agbo et al. 2021). As a result, air quality monitoring networks remain weak and limited, making assessment of exposure and health burden challenging. Priority is often given to other important issues, such as water supply, sanitation, malaria transmission and infectious disease control, for which there is sufficient evidence of public health relevance. Thus, it becomes extremely difficult to persuade governments to invest in air quality monitoring and other pollution control measures.

Climate Change and Air Pollution

The major driver linking the relationship between climate change and adverse health effects is fossil fuel combustion (Webb et al. 2023). This effect also underlines the interconnectedness between the two phenomena of air pollution and climate change, as the emissions driving both phenomena largely originate from the same source, i.e., fossil fuel or biofuel combustion (Fuller et al. 2022).

In Africa, one of the continents is most vulnerable to climate change (Moyo et al. 2023), increased incidence through fossil fuel combustion poses significant health risks, leading to various respiratory and cardiovascular diseases, injuries and premature deaths, especially in vulnerable populations (Wright et al. 2024).

The products of combustion—fine and ultrafine particles (for example, $PM_{2.5}$), long-lived greenhouse gases and short-lived climate pollutants (SLCPs), concomitantly cause air pollutants and climate warming. For example, haze formation, which is viewed as an air pollution phenomenon, can be controlled by local weather conditions (Yang et al. 2022). The primary SLCPs are methane, black carbon (i.e., soot), and hydrofluorocarbons (HFCs) (IPCC 2021). Methane emissions are one of the main precursors of ground-level ozone, and a major source of premature death (see: Fuller et al. 2022). Increase in air pollution may also lead to changes in production. Increased air pollution may also alter local and regional allergen production, with altered patterns depending on bio-climatic parameters (Singh and Kumar 2022). Disruptions in weather patterns also affect air quality by increasing and distributing air pollutants such as ground-level ozone, fine particles, wildfire smoke and dust.

These examples illustrate how adverse climate change scenarios can be directly or indirectly affected by aerosol concentrations as well as their optical properties; resulting not only in damage to earth systems but also to human health (Yang et al. 2022).

Addressing Issues

Environmental pollution is an intricate issue. The solution would require an intersectoral, multifaceted approach that will empower government policymakers, municipal officials and other stakeholders to combat pollution issues in a responsible and environmentally sound manner (Wang et al. 2024b).

In terms of Sub-Saharan Africa's pollution mitigation and soil environment cleanup, the prescriptions given by Tindwa and Singh in 2023 and those of Mesele et al. (2025), are instructive. The proffer given by Katoto et al. in 2019 to solve the problem of air pollution engenders different perspectives on problems and creates broader research questions; and can result in the framing of consensus clinical definitions and guidelines for establishing comprehensive cleanup plans.

Warren-Vega et al. (2023) noted the efforts of several research groups that have focussed on proposals for alternative solutions based on three fundamental aspects, namely, detection of contaminants, assessment of risks to public and environmental health and proposals to address the issues through remediation methods for soil, water and air.

The United Nations Environment Program (UNEP) is collaborating with governments, business and civil society to address pollution issues through the "Implementation Plan towards a Pollution-Free Planet" (UNEP 2017, 2019, 2023). This agenda, presented at the "Fourth United Nations Environment Assembly", identified five key action areas to address the gaps and challenges associated with pollution, namely: "knowledge, implementation, infrastructure, awareness and leadership".

The "Lancet Commission on Pollution and Health" (LCPH) addressed the full health and economic costs of air, water and soil pollution in 2017 (Landrigan et al. 2017); and through analysis of existing and emerging data, serious and under-reported contributions of pollution to GBD were revealed study (GBD 2019 Chronic Respiratory Diseases Collaborator 2023). The report revealed the economic costs of pollution in low-income and middle-income countries. In the latest version of this report, Fuller et al. (2022) used data from the latest iteration of the GBD study to calculate the health and economic costs of pollution up to 2019, identify areas where progress has been made, and outlined areas where further research is desirable. Finally, this study (Fuller et al. 2022) outlined a strategy for a joint and truly global policy response.

The most cited measures to address soil pollution problems in Africa to date (2025) include integrated approaches such as conservation agriculture, crop rotation, agroforestry, afforestation, organic farming and community participation (see, e.g., Masele et al. 2025).

Many research groups are actively working on ways to tackle water pollution problems in Africa. These initiatives focus on promoting water conservation and improving water quality by enhancing sanitation; and this includes investing in water treatment infrastructure, adopting water reuse strategies and educating communities about water conservation and the dangers of pollution. Ejiohuo et al. (2025) present a comprehensive framework for reducing the risks of water pollution, protecting biodiversity and promoting public health. LEMA's (2025) review offers sustainable solutions to the water pollution challenges affecting major rivers flowing through urban areas in East Africa.

Recent research on treatment technologies for AMD continues to increase (e.g., see: Rambabu et al., 2020; Nguégang and Embuche 2025), which highlights the devastating impacts on aquatic ecosystems. Yang et al. (2023) have examined and analysed the principles, advantages, disadvantages and future developments of various AMD processing methods. In Africa, most recent research on AMD treatment technologies has been conducted in South Africa (e.g., Masindi et al. 2018; Baloyi et al. 2024; Moghashane et al. 2025).

We still do not know enough about the negative health effects of air pollution in African cities. More effort is needed to research unique circumstances and diverse sources, as well as to establish efficient monitoring systems and incorporate effective and realistic enforcement mechanisms [for example, Okello et al. (2023); Capitanio et al. (2024)]. More research effort is needed to monitor ambient air pollution (AAP) in African cities, identify its main sources and reduce adverse health effects.

The relatively new technologies to combat environmental pollution such as "in situ air and steam sparging" (for remediation of soils) (see, e.g., Liang et al. 2024); phytoremediation (using plants to heal soil and water) (see, e.g., Kafle et al. 2022; Malunguza and Paschal 2024), nanotechnology-based water purification systems (see, e.g., Dhanda and Kumar 2023) and photocatalytic water purification (see, e.g., Omar et al. 2024), *all of which deserve further research.*

Moyo et al. (2023) set out a comprehensive approach aimed at addressing climate-sensitive health risks in Africa. Among these authors' prescriptions for tackling risks such as heat waves, vector-borne diseases and food insecurity, are the building of robust health care systems, implementation of climate-resilient infrastructure, promotion of sustainable practices across sectors and collaboration of stakeholders globally (Moyo et al. 2023).

Conclusion

Environmental pollution is probably the biggest cause of disease and death in Africa today (2025). Yet, this type of pollution and its harmful effects on people and ecosystems have not received due attention by governments in the region; and indeed, the regional policy response is still disappointing. Quantitative epidemiological data are lacking. Nor do we yet have fool-proof methods for determining risk and outcome. *More primary studies are needed to fill these knowledge and methodological gaps.*

References

Agbo KE, Walgraeve C, Eze JI, Ugwoke PE, Ukoha PO, Van Langenhove HR (2021) A review on ambient and indoor air pollution status in Africa. Atmos Pollut Res 12(2):243–260. https://doi.org/10.1016/j.apr.2020.11.006. Last accessed 11 Jan 2025

Ashong GW, Ababio BA, Kwaansa-Ansah EE, Koranteng SK, Muktar GD (2024) Investigation of fluoride concentrations, water quality, and non-carcinogenic health risks of borehole water in Bongo district, northern Ghana. Heliyon 10(6):e27554. https://doi.org/10.1016/j.heliyon.2024.e27554. Last accessed 21 Oct 2024

Baloyi J, Ramdhani N, Mbhele R, Simate GS (2024) Acid mine drainage from gold mining in South Africa: remediation, reuse, and resource recovery. Mine Water Environ 43:418–430. https://doi.org/10.1007/s10230-024-00994-2. Last accessed 24 Aug 2024

Borm PJA (2023) Talc inhalation in rats and humans: a review and appraisal of available evidence. J Occup Environ Med 65(2):152–159. https://doi.org/10.1097/JOM.0000000000002702. Last accessed 02 Aug 2023

Boudoire G, Giuffrida G, Liuzzo M, Bobrowski N, Calabrese S (2022) Chemical variability in volcanic gas plumes and fumaroles along the East African Rift system: new insights from the Western Branch. Chem Geol 596:120811. https://doi.org/10.1016/j.chemgeo.2022.120811. Last accessed 29 Sept 2024

Bredeck G, Dobner J, Rossi A, Schins RP (2024) Saharan dust induces the lung disease-related cytokines granulocyte-macrophage colony-stimulating factor and granulocyte colony-stimulating factor. Environ Int 186:108580. https://doi.org/10.1016/j.envint.2024.108580. Last accessed 26 Sept 2024

Bruce M, Limin M (2021) Recent advances on water pollution research in Africa: a critical review. Int J Sci Adv (IJSCIA) 2(3):371–374. https://www.ijscia.com/wp-content/uploads/2021/06/Volume2-Issue3-May-Jun-No.96-375-386.pdf. Last accessed 12 Aug 2025

Capitanio L, Ratte S, Gautier S, Josseran L (2024) Impact of air pollution on mortality: geo-epidemiological study in French-speaking Africa. Heliyon 10(20):e39473. https://doi.org/10.1016/j.heliyon.2024.e39473. Last accessed 12 Aug 2025

Chakraborty A, Biswas T, Sarkar P, Haque M (2025) A mini review on tuberculosis risk in coal mining: a matter of exposure or environment? Int J Pharm Pharm Res (IJPPR) 31(8). 3.Dr_.-Ayeshik-Chakraborty1-Dr.-Tanisha-Biswas2-Priyanka-Sarkar3-Dr.-Morziul-Haque4.pdf. Last accessed 24 Aug 2025

Combatting toxic chemical elements pollution for Sub-Saharan Africa's ecological health. Environ Pollut Manag 2:42–62. https://doi.org/10.1016/j.epm.2025.01.003. Last accessed 12 Aug 2025

Davies TC (2023) Current status of research and gaps in knowledge of geophagic practices in Africa. Front Nutr 9:1084589. https://www.ncbi.nlm.nih.gov/pmc/articles/PMC9987423/. Last accessed 27 Feb 2024

Demelash H, Beyene A, Abebe Z, Melese A (2019) Fluoride concentration in ground water and prevalence of dental fluorosis in Ethiopian Rift Valley: systematic review and meta-analysis. BMC Public Health 19:1–9. https://doi.org/10.1186/s12889-019-7646-8. Last accessed 21 Oct 2024

Dhanda N, Kumar S (2023) Smart and innovative nanotechnology applications for water purification. Hybrid Adv 3(3):100044. https://doi.org/10.1016/j.hybadv.2023.100044. Last accessed 29 Sept 2024

Ejiohuo O, Onyeaka H, Akinsemolu A, Nwabor OF, Siyanbola KF, Tamasiga P, Al-Sharify ZT (2025) Ensuring water purity: Mitigating environmental risks and safeguarding human health. Water Biol Secur 4(2):100341. https://doi.org/10.1016/j.watbs.2024.100341. Last accessed 12 Aug 2025

Ermolin MS, Ivaneev AI, Fedyunina NN, Fedotov PS (2021) Nanospeciation of metals and metalloids in volcanic ash using single particle inductively coupled plasma mass spectrometry. Chemosphere 281:130950. https://doi.org/10.1016/j.chemosphere.2021.130950. Last accessed 29 Sept 2024

FAO and UNEP (United Nations Food and Agricultural Organisation and United Nations Environment Programme) (2021) Global assessment of soil pollution: Report. Rome. ISBN: 978-92-5-134469-9. https://doi.org/10.4060/cb4894en. Last accessed 21 Jun 2023

Fayiga AO, Ipinmoroti MO, Chirenje T (2018) Environmental pollution in Africa. Environ Dev Sustain 20:41–73. https://doi.org/10.1007/s10668-016-9894-4. Last accessed 22 Jun 2023

Fisher S, Bellinger DC, Cropper ML, Kumar P, Binagwaho A, Koudenoukpo JB, Park Y, Taghian G, Landrigan PJ (2021) The Lancet Planetary Health 5(10):e681–e688. https://doi.org/10.1016/S2542-5196(21)00201-1. Last accessed 17 Jun 2023

Fubini B, Fenoglio I (2007) The toxic potential of mineral dust. ELEMENTS 3:407–414. https://www.elementsmagazine.org/wp-content/uploads/archivearticles/e3_6/fubini.pdf. Last accessed 02 Aug 2023

Fuller CH, Amegah AK (2022) Limited air pollution research on the African continent: time to fill the gap. Int J Environ Res Public Health 19(11):6359. https://doi.org/10.3390/ijerph19116359. Last accessed 11 Jan 2025

Fuller R, Landrigan PJ, Balakrishnan K, Bathan G, Bose-O'Reilly S, Brauer M, Caravanos J, Chiles T, Cohen A et al (2022) Pollution and health: a progress update. Lancet Planet Health 6(6):e535–e547. https://doi.org/10.1016/S2542-5196(22)00090-0. Epub 2022 May 18. Erratum in: Lancet Planet Health. 14. https://www.thelancet.com/journals/lanplh/article/PIIS2542-5196(22)00090-0/fulltext. Last accessed 31 Jul 2023

GAN (Global Africa Network) (2024) Digitalisation in Mining Africa. https://www.globalafricanetwork.com/mining/digitalisation-in-mining-africa-2024/. Last accessed 29 Sept 2024

GBD 2019 Chronic Respiratory Diseases Collaborators (2023) Global Burden of Chronic Respiratory Diseases and Risk Factors 1990–2019: An update from the global burden of disease study 2019. EClinicalMedicine (Part of the Lancet Discovery Science) 59:101936. https://doi.org/10.1016/j.eclinm.2023.101936. Last accessed 13 Aug 2023

Gbenga Wilfred A, Ohonga A (2024) The effects of fossil fuel consumption-related CO_2 on health Outcomes in South Africa. Sustainability 16(11):4751. https://doi.org/10.3390/su16114751. Last accessed 26 Sept 2024

Harr B, Pu B, Jin Q (2024) The emission, transport, and impacts of the extreme Saharan dust storm of 2015. Atmos Chem Phys 24:8625–8651. https://doi.org/10.5194/acp-24-8625-2024. Last accessed 26 Sept 2024

HEI (Health Effects Institute) (2022) The state of air quality and health impacts in Africa. A report from the state of global air initiative. Health Effects Institute, Boston, MA. ISSN 2578-6881. https://www.stateofglobalair.org/sites/default/files/documents/2022-10/soga-africa-report.pdf. Last accessed 30 Jun 2023

HEI/IHME (Health Effects Institute/Institute for Health Metrics and Evaluation) (2024) State of Global Air (SoGA) Report https://www.stateofglobalair.org/resources/report/state-global-air-report-2024. Last accessed 24 Sept 2024

Huang X, Rom X, Gordon T, Finkelman RB (2006) Interaction of iron and calcium minerals in coals and their roles in coal dust-induced health and environmental problems. Rev Mineral Geochem 64(1):153–178. https://doi.org/10.2138/rmg.2006.64.6. Last accessed 02 Aug 2023

Igboama WN, Hammed OS, Fatoba JO, Aroyehun MT, Ehiabhili (2022) Review article on impact of groundwater contamination due to dumpsites using geophysical and physiochemical methods. Appl Water Sci 12:130. https://doi.org/10.1007/s13201-022-01653-z. Last accessed 25 Jun 2023

Ighalo JO, Adeniyi AG, Adeniran AJ, Ogunniyi (2021) A systematic literature analysis of the nature and regional distribution of water pollution sources in Nigeria. J Clean Prod 283. 124566. ISSN: 0959–6526. (1) A systematic literature analysis of the nature and regional distribution of water pollution sources in Nigeria | Request PDF. Last accessed 21 Oct 2024

IPCC (The Intergovernmental Panel on Climate Change) (2021) Climate Change 2021: the physical science basis—contribution of working group i to the sixth assessment report of the intergovernmental panel on climate change. Cambridge University Press. https://www.ipcc.ch/report/ar6/wg1/#FullReport. Last accessed 02 Aug 2023

Irunde R, Ijumulana J, Ligate IF, Maity JP, Ahmad AA, Mtamba J, Mtalo F, Bhattacharya P (2022) Arsenic in Africa: potential sources, spatial variability, and the state of the art for arsenic removal using locally available materials. Groundw Sustain Dev 18:100746. https://doi.org/10.1016/j.gsd.2022.100746

Kafle A, Timilsina A, Gautam A, Adhikari K, Bhattarai A, Aryal N (2022) Phytoremediation: Mechanisms, plant selection and enhancement by natural and synthetic agents. Environ Adv 8:100203. https://doi.org/10.1016/j.envadv.2022.100203. Last accessed 12 Aug 2025

Kapwata T, Wright CY, Reddy T, Street R, Kunene Z, Mathee A (2023) Relations between personal exposure to elevated concentrations of arsenic in water and soil and blood arsenic levels amongst people living in rural areas in Limpopo, South Africa. Environ Sci Pollut Res 30:65204–65216. https://doi.org/10.1007/s11356-023-26813-9. Last accessed 21 Oct 2024

Katoto PDMC, Byamungu L, Brand AS, Mokaya J, Strijdom H, Goswami N, De Boever P, Nawrot TS, Nemery B (2019) Ambient air pollution and health in Sub-Saharan Africa: current evidence, perspectives and a call to action. Environ Res 173:174–188. https://doi.org/10.1016/j.envres.2019.03.029(Accessed18.06.2023)

King DCP, Watts MJ, Hamilton EM, Mortimer RJG, Coffey M, Osano O, Ondayo MA, Di Bonito M (2024) Field method for preservation of total mercury in waters, including those associated with artisanal scale gold mining. Anal Methods: Adv Methods Appl 16(17):2669–2677. https://doi.org/10.1039/d3ay02216a. Last accessed 21 Oct 2024

Kirk K (2023) Aerosols: small particles with big climate effects. NASA Global Climate Change. https://climate.nasa.gov/explore/ask-nasa-climate/3271/aerosols-small-particles-with-big-climate-effects/. Last accessed 02 Aug 2023

Kotsyfakis M, Zarogiannis S, Patelarou E (2019) The health impact of Saharan dust exposure. Int J Occup Med Environ Health 32(6):749–760. https://doi.org/10.13075/ijomeh.1896.01466. Last accessed 26 Dec 2022

Laker MC (2023) Environmental impacts of gold mining—With special reference to South Africa. Mining 3:205–220. https://doi.org/10.3390/mining3020012. Last accessed 30 Jun 2023

Landrigan PJ, Fuller Acosta NJR, Adeyi O, Arnold R, Baldé AB et al (2017) The lancet commission on pollution and health. The Lancet 391:10119. Last accessed 26 Sept 2024

Lelieveld J, Haines A, Burnett R, Tonne C, Klingmüller K, Münzel T, Pozzer A (2023) Air pollution deaths attributable to fossil fuels: observational and modelling study. BMJ (Clinical Research Ed.) 383:e077784. https://doi.org/10.1136/bmj-2023-077784. Last accessed 26 Sept 2024

Lema MW (2025) Contamination of urban waterways: a mini review of water pollution in the rivers of East Africa's major cities. HydroResearch 8:307–315. https://doi.org/10.1016/j.hydres.2024.11.004. Last accessed 12 Aug 2025

Liang B, Zhang X, Wu Z, Tang L, Chen X (2024). Characterization of air distribution during horizontal well air sparging with various sparging tube configurations. J Clean Prod 471:143402. https://doi.org/10.1016/j.jclepro.2024.143402. Last accessed 29 Sept 2024

Lin L, Yang H, Xu X (2022) Effects of water pollution on human health and disease heterogeneity: a review. Front Environ Sci 10. Sec.: Water and Wastewater Management. https://doi.org/10.3389/fenvs.2022.880246

Lucero-Prisno III DE, Shomuyiwa DO, Kouwenhoven MBN et al (2023). Top 10 public health challenges to track in 2023: Shifting focus beyond a global pandemic. Public Health Chall 2:e86. https://doi.org/10.1002/puh2.86. Last accessed 20 Oct 2023

Malunguja GK, Paschal M (2024) Evaluating potential phytoremediators to combat detrimental impacts of mining on biodiversity: a review focused in Africa. Discov Environ 2:94. https://doi.org/10.1007/s44274-024-00133-2. Last accessed 29 Sept 2024

Marwa J, Lufingo M, Noubactep C, Machunda R (2018) Defeating fluorosis in the East African rift valley: transforming the Kilimanjaro Into a rainwater harvesting park. Sustainability. MDPI 10(11):1–12. https://doi.org/10.3390/su10114194. Last accessed 29 Sept 2024

Masindi V, Chatzisymeon E, Kortidis I, Foteinis S (2018) Assessing the sustainability of acid mine drainage (AMD) treatment in South Africa. Sci Total Environ 635, 793–802. https://doi.org/10.1016/j.scitotenv.2018.04.108. Last accessed 24 Aug 2025

Mbuende M (2022) Determination of natural radioactivity in the soil of Omaruru, Namibia (Doctoral dissertation, University of Namibia). https://repository.unam.edu.na/items/7658d08d-81. Last accessed 30 Dec 2025

Mesele SA, Mechri M, Okon MA, Isimikalu TO, Wassif OM, Asamoah E, Khurshid C (2025) Current problems leading to soil degradation in Africa: Raising awareness and finding potential solutions. Eur J Soil Sci 76(1):e70069, 1–22. https://doi.org/10.1111/ejss.70069. Last accessed 12 Aug 2025

Mogashane TM, Maree JP, Mokoena L, Tshilongo J (2025) Research activities on acid mine drainage treatment in South Africa (1998–2025): trends, challenges, bibliometric analysis and future directions. Water 17:2286. https://doi.org/10.3390/w17152286. Last accessed 24 Aug 2025

Mojakwana A, Ewuoso C (2023) An Ubuntu-based reflection on the public health impact of silica dust exposure in the South African mining industry. J Glob Health Rep 7:e2023028. https://doi.org/10.29392/001c.77498

Moyo E, Nhari LG, Moyo P, Murewanhema G, Dzinamarira T (2023) Health effects of climate change in Africa: a call for an improved implementation of prevention measures. Eco-Environ Health 2(2):74–78. https://doi.org/10.1016/j.eehl.2023.04.004. Last accessed 13 Aug 2025

Mphaga KV, Utembe W, Rathebe PC (2024) Radon exposure risks among residents proximal to gold mine tailings in Gauteng Province, South Africa: a cross-sectional preliminary study protocol. Front Public Health 12:1328955. https://doi.org/10.3389/fpubh.2024.1328955. Last accessed 21 Oct 2024

Münzel T, Hahad O, Daiber A, Landrigan PJ (2023) Soil and water pollution and human health: what should cardiologists worry about? Cardiovasc Res 31 119(2):440–449. https://doi.org/10.1093/cvr/cvac082

Musiige D, Mundike J, Makondo C (2024) Evaluation of potentially toxic metals in tailings from Busia gold mine fields of eastern Uganda. J Clean Prod 469:143222. https://doi.org/10.1016/j.jclepro.2024.143222. Last accessed 19 Sept 2024

Nemery B (2022) Metals and the respiratory tract. In: Nordberg GF, Costa M (eds) Handbook on the toxicology of metals (fifth edn), vol 1, chapter 19: general considerations, pp 421–443. https://doi.org/10.1016/B978-0-12-823292-7.00030-9. Last accessed 02 Aug 2023

Nguegang B, Ambushe AA (2025) Sustainable acid mine drainage treatment: a comprehensive review of passive, combined, and emerging technologies. Environ Eng Res 30(4):Article 240592. https://doi.org/10.4491/eer.2024.592. Last accessed 24 Aug 2025

Nordstrom DK (2022) Fluoride in thermal and non-thermal groundwater: Insights from geochemical modeling. Sci Total Environ 824:153606. https://doi.org/10.1016/j.scitotenv.2022.153606. Last accessed 25 Sept 2024

Ogundele LT, Owoade OK, Hopke PK, Olise FS (2017) Heavy metals in industrially emitted particulate matter in Ile-Ife, Nigeria. Environ Res 156:320–325. https://doi.org/10.1016/j.envres.2017.03.051 PMID: 28390299

Okello G, Nantanda R, Awokola B, Thondoo M, Okure D, Tatah L, Bainomugisha E, Oni T (2023) Air quality management strategies in Africa: a scoping review of the content, context, co-benefits and unintended consequences. Environ Int 171:107709. https://doi.org/10.1016/j.envint.2022. 107709. Last accessed 09 Jan 2025

Oliver MA, Gregory PJ (2015) Soil, food security and human health: a review: soil, food security and human health. Eur J Soil Sci 66(2):257–276. https://doi.org/10.1111/ejss.12216. Last accessed 19 Aug 2023

Omar PJ, Tripathi R, Azamathulla H (2024) Photocatalytic water purification technology for contaminated water treatment. Top Catal 67:959–960. https://doi.org/10.1007/s11244-024-019 91-z. Last accessed 29 Sept 2024

Philippova A, Aringazina R, Kurmanalina G, Beketov V (2022) Epidemiology, clinical and physiological manifestations of dust lung disease in major industrial centers. Emerg Themes Epidemiol 19(1):3. https://doi.org/10.1186/s12982-022-00111-0. Last accessed 29 Sept 2024

Pouri N, Karimi B, Kolivand A, Mirhoseini SH (2024) Ambient dust pollution with all-cause, cardiovascular and respiratory mortality: a systematic review and meta-analysis. Sci Total Environ 912:168945. https://doi.org/10.1016/j.scitotenv.2023.168945. Last accessed 26 Sept 2024

Raimi MO, Iyingiala AA, Sawyerr OH, Saliu AO, Ebuete AW et al (2022) Leaving no one behind: impact of soil pollution on biodiversity in the global south: a global call for action. In: Chibueze Izah S (eds) Biodiversity in africa: potentials, threats and conservation. sustainable development and biodiversity 29. Springer, Singapore. https://doi.org/10.1007/978-981-19-3326-4_8. Last accessed 21 Jun 2023

Rambabu K, Banat F, Pham QM, Ho SH, Ren NQ, Show PL (2020) Biological remediation of acid mine drainage: Review of past trends and current outlook. Environ Sci Ecotechnology 2:100024. https://doi.org/10.1016/j.ese.2020.100024. Last accessed 24 Aug 2025

Rathebe PC, Khosi L, Kholopo M (2025) Indoor concentration of radon in residential houses proximal to gold mine tailings—A review of sub-Saharan Africa studies. Environ Forensics 1–12. https://doi.org/10.1080/15275922.2024.2431325. Last accessed 12 Aug 2025

Ripanda A, Hossein M, Rwiza MJ, Nyanza EC, Selemani JR, Nkrumah S, Bakari R, Alfred MS, Machunda RL, Vuai SAH et al (2025)

Sakhamuri S, Cummings S (2019) Increasing trans-Atlantic intrusion of Sahara dust: a cause of concern? The Lancet. Planet Health 3(6):e242–e243. https://doi.org/10.1016/S2542-519 6(19)30088-9. Last accessed 26 Sept 2024

Schiavo B, Morton-Bermea O, Meza-Figueroa D, Acosta-Elías M, González-Grijalva B, Armienta-Hernández MA, Inguaggiato C, Valera-Fernández D (2023) Characterization and polydispersity of volcanic ash nanoparticles in synthetic lung fluid. Toxics 11(7): 624. https://doi.org/10.3390/ toxics11070624. Last accessed 26 Sept 2024

SDWF (Safe Drinking Water Foundation) (Updated 2023) Mining and water pollution. Safe drinking water foundation. https://www.safewater.org/fact-sheets-1/2017/1/23/miningandwat erpollution. Last accessed 13 Sept 2025

Shaji E, Sarath KV, Santosh M, Krishnaprasad PK, Arya BK, Babu, Manisha S (2024) Fluoride contamination in groundwater: a global review of the status, processes, challenges, and remedial measures. Geosci Front 15(2):101734. https://doi.org/10.1016/j.gsf.2023.101734. Last accessed 30 Sept 2024

Singh N, Irusen EM (2025) Coal mining as a risk factor for tuberculosis—commodity or circumstance? South Afr J Infect Dis 40(1):a708. https://doi.org/10.4102/sajid.v40i1.708. Last accessed 24 Aug 2025

Singh AB, Kumar P (2022) Climate change and allergic diseases: an overview. Front Allergy 3:964987. https://doi.org/10.3389/falgy.2022.964987. Last accessed 02 Aug 2023

Sonwani S, Madaan S, Arora J, Suryanarayan S, Rangra D, Mongia N, Vats T, Saxena P (2021) Inhalation exposure to atmospheric nanoparticles and its associated impacts on human health: a review. Front Sustain Cities 3:2021. https://doi.org/10.3389/frsc.2021.690444

Tindwa HJ, Singh BR (2023) Soil pollution and agriculture in sub-Saharan Africa: state of the knowledge and remediation technologies. Front Soil Sci 2. https://doi.org/10.3389/fsoil.2022.1101944

UNDRR (United Nations Office of Disaster Reduction) (2024) Volcanic gases and aerosols. Unique identifier/Notation GH0016. https://www.undrr.org/understanding-disaster-risk/terminology/hips/gh0016#:~:text=Health%20impacts%3A%20The%20common%20effects,and%20(iii)%20asthma%20aggravation. Last accessed 26 Sept 2024

UNEP (United Nations Environment Programme) (2017) Towards a pollution-free planet background report. United Nations environment programme, Nairobi, Kenya. https://wedocs.unep.org/bitstream/handle/20.500.11822/21800/UNEA_towardspollution_long%20version_Web.pdf?sequence=1&isAllowed=y. Last accessed 26 Sept 2024

UNEP (United Nations Environment Programme) (2019) Implementation plan "towards a pollution-free planet": resolution/adopted by the United Nations environment assembly. UNEP Environment Assembly (4th Session: 2019). https://digitallibrary.un.org/record/3982547?ln=en&v=pdf. Last accessed 26 Sept 2024

UNEP (United Nations Environment Programme) (2023) The United Nations system: common approach towards a pollution-free planet. https://unemg.org/wp-content/uploads/2023/12/UNEP_EMG_UN-system-common-approach-to-Pollution-231122.pdf. Last accessed 26 Sept 2024

Usman US, Salh YHM, Yan B, Namahoro JP, Zeng Q, Sallah I (2024) Fluoride contamination in African groundwater: Predictive modelling using stacking ensemble techniques. Sci Total Environ 957:177693. https://doi.org/10.1016/j.scitotenv.2024.177693. Last accessed 12 Aug 2025

Wang L, Xie J, Hu Y, Tian Y (2022) Air pollution and risk of chronic obstructed pulmonary disease: the modifying effect of genetic susceptibility and lifestyle. EBioMedicine, Part Lancet Discov Sci 79:103994. https://doi.org/10.1016/j.ebiom.2022.103994. Last accessed 23 Jul 2021

Wang Z, Zhang J, Yang Y, Cao M, Ma J, Li S, Shao H, Du Z (2024a) Current status, trends, and predictions in the burden of coal worker's pneumoconiosis in 204 countries and territories from 1990 to 2019. 10(19):e37940. https://doi.org/10.1016/j.heliyon.2024.e37940. Last accessed 26 Sept 2024

Wang F, Xiang L, Sze-Yin Leung K, Elsner M, Zhang Y, Guo Y, Pan B, Sun H, An T, Tiedje JM (2024b) Emerging contaminants: a one health perspective. The Innovation [(Cambridge (Mass.)] 5(4):100612. https://doi.org/10.1016/j.xinn.2024.100612. Last accessed 07 Sept 2025

Warren-Vega WM, Campos-Rodríguez A, Zárate-Guzmán AI, Romero-Cano LA (2023) A current review of water pollutants in American continent: trends and perspectives in detection, health risks, and treatment technologies. Int J Environ Res Public Health 20(5):4499. https://doi.org/10.3390/ijerph20054499. Last accessed 26 Sept 2024

Webb D, Hanssen ON, Marten R (2023) The health sector and fiscal policies of fossil fuels: an essential alignment for the health and climate change agenda. BMJ global health 8(Suppl 8):e012938. https://doi.org/10.1136/bmjgh-2023-012938. Last accessed 30 Sept 2024

Wright CY, Kapwata T, Naidoo N, Asante KP, Arku RE, Cissé G, Simane B, Atuyambe L, Berhane K (2024) Climate change and human health in Africa in relation to opportunities to strengthen mitigating potential and adaptive capacity: strategies to inform an African "brains trust". Ann Glob Health 90(1):7. https://doi.org/10.5334/aogh.4260. Last accessed 13 Aug 2025

Xin X, Shentu J, Zhang T, Yang X, Baligar VC, He Z (2022) Sources, indicators, and assessment of soil contamination by potentially toxic metals. Sustainability 14(23):15878 (2022). https://doi.org/10.3390/su142315878. Last accessed 20 Jun 2023

Yang Y, Ho HC, Lolli SL, Yang X (2022) Editorial: climate change, aerosol pollution and public health risk in an urban context. Front Climate Clim Serv 4. https://doi.org/10.3389/fclim.2022.862982. Last accessed 25 Jun 2023

Yang Y, Li, Li T, Liu PU, Zhang B, Che L (2023) A review of treatment technologies for acid mine drainage and sustainability assessment. J Water Process Eng 55, 104213. https://doi.org/10.1016/j.jwpe.2023.104213. Last accessed 24 Aug 2025

Yi X, He Y, Zhang Y, Luo Q, Deng C, Tang G, Zhang J, Zhou X, Luo H (2023) Current status, trends, and predictions in the burden of silicosis in 204 countries and territories from 1990 to 2019. Front Public Health 11:1216924. https://doi.org/10.3389/fpubh.2023.1216924. Last accessed 26 Sept 2024

Chapter 3
Criticality of the Optimal Intake Level of Nutritional and Potentially Toxic Elements in the African Food Chain

T. C. Davies and X. M. Mkhize

Abstract There has been an upsurge in Medical Geology research in Africa during the last decade, driven by the increased realisation of the criticality of the optimal intake levels of nutritional elements and the need for elimination of potentially toxic elements PTEs) from the food chain for maintaining health. The studies have focussed on understanding the geochemical fluxes of both nutritional and PTEs in the groundwater-soil-food crop continuum, that largely define the nutritional quality of the African diet, and hence the correct balance of micronutrients and vitamins in metabolic processes. Geochemical processes carried to extremes in humid and sub-humid sectors of Africa's surface environment have led to the evolution of large areas of the Continent that are deficient in some key nutritional elements and other areas that have high amounts of PTEs. We highlight in this review, studies that speak to the criticality of the intake levels of nutritional elements; in particular, the amount that reaches the metabolic centres of activity, which in turn, is determined by *geochemobiological* parameters such as bioavailability, bioaccessibility and *dose-reponse relationships*. It is this amount of the element (nutritional or PTE) that determines whether nutritional benefit is realised, or disease is produced. Gaps in knowledge (*italicised*) regarding optimal intake levels of certain nutritional elements have been identified, illnesses due to excessive intake levels of PTEs are referred, and ways of obviating the resulting maladies are predicated.

Introduction

The trend in Medical Geology research in Africa during the last decade has been towards determining the circulation of potentially toxic elements (PTEs) in the water-soil-food crop nexus that enters the food chain (see, e.g., Fayiga et al. 2018; Kihara

T. C. Davies (✉) · X. M. Mkhize
Faculty of Applied and Health Sciences, Mangosuthu University of Technology, Umlazi, KwaZulu Natal Province, Republic of South Africa
e-mail: daviestheophilus2025@yahoo.com

T. C. Davies (ed.), *Recent Advances in Medical Geology Research in Africa: A Decadal View*, SpringerBriefs in Earth System Sciences,
https://doi.org/10.1007/978-3-032-13754-8_3

et al. 2020; Nana et al. 2023; Adhikari and Struwig 2024; Boahen et al. 2024; John et al. 2024; Gaothobogwe et al. 2025; Ripanda et al. 2025; Siddig et al. 2025; and references within). The prime motivator of this approach has been the increasing realisation of the significance of the entry—largely through the diet—of varying concentration levels of elements that may be bioavailable for negative interactions in metabolic processes that produce diseases, some of whose diagnoses are still ill-defined (Davies 2022).

The surface geochemistry of Africa is unique and extremely complex. Separation, remobilisation and redistribution of elements by surficial geogenic processes (such as intense tropical weathering, leaching and lateritisation) have created large areas exhibiting deficiency in some key nutritional elements, and excessive amounts in others (see: Davies 2010). Similarly, large amounts of PTEs are dispersed widely in certain arable soils, especially those associated with mineralised terrains. Examples of public health conditions that may reflect nutrient poor environments in Africa include iodine deficiency disorders (IDD), symptoms of fluoride deficiency, the *geophagy phenomenon* (deliberate or sometimes inadvertent consumption of earth materials) as a possible iron or other essential element supplementer, selenium deficiency and HIV-AIDS and iron deficiency.

This Chapter summarises the results from recent researches on variations in the amount of element that enters the body, and the actual fraction that takes part in metabolic reactions to produce various effects (beneficial or deleterious). The role of the geochemobiological properties that determine this fraction, stemming from the environmental media (soil, water, food crops or air) to the metabolic pathways—bioavailability, bioaccessibility and dose–response relationships—is discussed, using the nutritional circulation and deficiency of iodine, fluorine and selenium, three geomedically important elements that are sometimes deficient in the African diet; and the phenomenon of *geophagy* (the deliberate or sometimes inadvertent consumption of earth materials).

Criticality of the Optimal Intake Level of Nutritional Elements and Excess of Potentially Toxic Elements in the African Food Chain

Sufficiency of intake levels of nutritional elements does not guarantee availability of the nutrient in the metabolic pathways. Dysfunctioning of the transport mechanism, the masking effect of competing ions or antagonistic interactions are all real possibilities. Adherence to the "recommended dietary allowance" (RDA) or "daily intake level" (DIL) for the nutritional elements (given for the examples discussed in this Chapter), though of high importance, does not in all cases result in the warding off of disease.

Resolutions 7 and 8 respectively, made at the first East and Southern Africa Regional Workshop on Geomedicine held in 1999 in Nairobi, Kenya (Davies 1999), noted the dearth in knowledge then (1999), on how to effectively monitor the flow of nutritional and PTEs in the environment that leads to human exposure through the consumption of food and water (Resolution 7); and the need for further research on *speciation, bioavailability*, as well as *bioaccessibility* of the trace elements in soils, plant materials, and in humans and animals (Resolution 8) (Davies 1999).

The essential trace elements function primarily as catalysts in enzyme systems and other metabolic processes. Both the essential elements and nonessential elements can potentially give rise to disease depending on the concentration levels (excess or deficiency) present in living tissues. People are exposed to PTEs in soil, water and dust during everyday activities. The PHEs include mercury, lead, arsenic, cadmium and polycyclic aromatic hydrocarbons (PAHs).

Today, in Africa, important advances have been made over the last decade of research on these issues, including an improvement in our mechanistic understanding of nutrient and PTE cycling. These developments have helped foster our ability to more precisely assess the nutritional efficacy or disease-causing potential of the different categories of elements.

Bioavailability, Bioaccessibility and the Dose–Response Relationships

A natural physiological response when the body takes up foreign substances is the production of fluids at the site of exposure to help dilute, solubilise, or physically clear the substances. *Bioavailability, bioaccessibility* and *dose–response relationships* are among the main determinants of whether or not adverse effects are to be expected when organisms or plants are exposed to PTEs (Peijnenburg and Jager 2003).

Toxicologists define "bioavailability" as the fraction of an administered dose of a substance (drug, chemical element, xenobiotic) that is absorbed via an exposure route, reaches the bloodstream, and is transported in the body to a site of toxicological action (Price and Patel 2023). The "bioaccessibility" of a substance is the fraction that can be dissolved by body fluids [i.e., in the gastrointestinal track (G-I tract) or in the lungs] and therefore, is available for absorption (Petruzzelli et al. 2020). High levels of PTEs are not necessarily indicative of adverse effects. These effects are related to bioavailable fractions, which are involved in plant uptake and transfer to the food chain (Newell et al. 2025).

Dose Response

It has already been established that many elements (the nutritional elements) are essential to the effective functioning of the myriad biochemical processes in the body, yet can become toxic after intense exposures over short periods of time (acute toxicity) or moderately high exposures over longer periods of time (chronic toxicity).

The threshold above which a material becomes toxic is a complex function of the substance, exposure route, chemical form of the substance as it is absorbed and presented at the site of action, and, to a lesser extent, the genetic makeup of the exposed individual (Arnot et al. 2022).

As the famous Renaissance physician Paracelsus (1493–1541) observed, "All substances are poisonous, there are none which is not a poison; the right dose is what differentiates a poison from a remedy" [quoted from Grandjean (2016)]. This frequently repeated quote underscores the long-recognised importance of understanding exposure intensity and duration (the dose) in assessing the health effects of potential toxicants. The concept of *"dose response"* states that the greater amount of a toxicant taken up by the organism, the greater the toxicological response (see, e.g., Tsatsakis et al. 2018; Farinde 2023; Fig. 3.1). However, the true determinant of toxicity is based on the concentration and form of the toxicant at the site of action. The body has a remarkable capability for clearing or diminishing the toxic effects of a wide variety of toxicants such as *heavy metals*, pesticides or asbestos fibres if the dose is small; however, toxicity results when the dose exceeds the body's inherent toxicant mitigation mechanisms (Moffett et al. 2022; Fig. 3.1).

The human body needs some 20 essential elements for it to function properly, and among them, are several metallic elements (Jomova et al. 2022). Trace elements are involved in vital biochemical processes. Some act as cofactors for many enzymes, as well as serve as centres for stabilising structures of enzymes and proteins (Prashanth et al. 2015; Islam et al. 2023). The relationship between essential trace elements and physiological functioning takes the form of a U-shaped curve. The points at which deficiencies and, in contrast, poisoning occur are specific to each element (Panzeri et al. 2024), with impairment occurring at levels of inadequate intake, optimal function occurring at intermediate levels, and toxicity occurring at levels of excessive

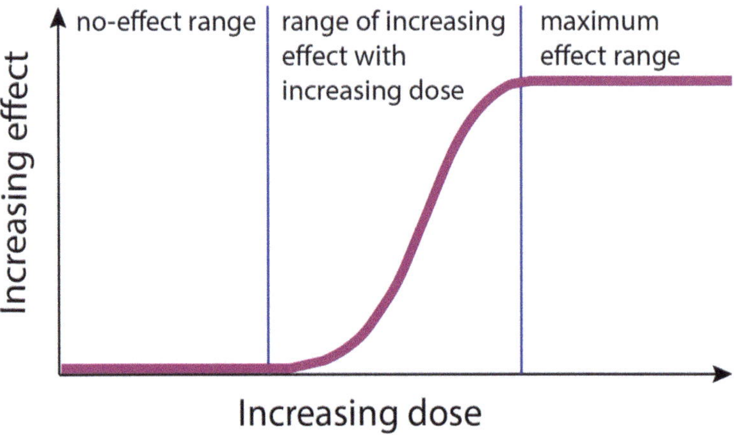

Fig. 3.1 Typical dose–response curve. Credit: Toxicology Education Foundation. https://toxedfoun dation.org/basics-of-dose-response/ (Accessed 01.02.2025). Reproduced with permission granted on 17.08.2025

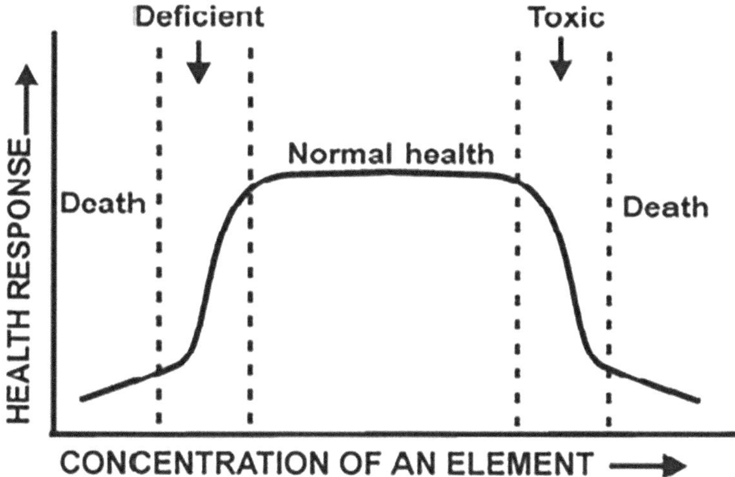

Fig. 3.2 A generalised dose–response curve demarcating regions of extreme deficiency and extreme toxicity. From: Bowman et al. (2003). Medical geology: New relevance in the earth sciences, EPISODES 26 (4), 125–133. https://www.sgb.gov.br/documents/d/guest/episodes_article-pdf#:~:text=The%20interdisciplinary%20field%20of%20%22Medical,be%20found%20throughout%20the%20world (Accessed 19.08.225). Reproduced with permission granted on 10.08.2025

consumption (Dictionary of Toxicology 2024). As the dose is increased to the point that the deficiency no longer exists, and no adverse response is therefore detected, the organism reaches a state of *homeostasis* (maintenance of a stable and balanced internal environment despite external changes).

At very low concentrations, there is a high-level adverse effect (death in extreme cases), which decreases as the dose increases; similarly, at very high doses, there is a high-level adverse effect (death in extreme cases) that decreases with decreasing dose (Fig. 3.2).

Examples of Health Issues Arising from the Evolution of Nutrient Poor Environments and Nutritional Element Imbalances

Kihara et al. (2020) found evidence of widespread but varying micronutrient deficiencies co-occurring in Sub-Saharan Africa (SSA) arable soils, with the predominance of zinc deficiency. In the same year (2020), Ibia et al. also noted that boron, iron, molybdenum and copper deficiencies are rife in SSA soils.

Iodine Deficiency Disorders

Iodine (I) is a key element used by the thyroid gland to make the hormone thyroxine (T4). Up to 80% of the body's iodine supply is stored in the thyroid, making this element a critically important nutrient (Zimmermann 2020). It is one of the first chemical elements to graphically illustrate the unequivocal link existing between the geochemical landscape of a region and the environmental health of its people. The element also exquisitely illustrates the criticality of the optimal intake dose for an essential element; for, in Africa, both iodine deficiency (ID) and excess are associated with adverse health effects (see, e.g., Kassim et al. 2012).

The DIL of iodine has for long been established as 140 μg (μg) of iodine a day as the adult requirement (NHS, Undated)]. It was then left to contemporary geochemists to determine the element's levels in the various landscapes (see: Hoseinzadeh and Taha 2024), making use of the opportunity to appositely demonstrate the power of multidisciplinarity in therapeutic endeavours (see: Box 1). This trace amount of iodine (140 μg/day) is essential for the correct functioning of the thyroid gland of adult mammals; and must be present and maintained throughout the life of the mammal by way of ingesting small quantities of the element from drinking water and various foodstuffs.

The absence of adequate amounts of iodine in the body leads to a spectrum of disorders collectively referred to as *iodine deficiency disorders* (IDDs), which include goitre, cretinism, low psychomotor performance, reduced cognitive function, still births, and many other sequelae (Saha et al. 2019).

Despite the apparent success of the Universal Salt Iodisation (USI) Programme, spear-headed by the World Health Organisation (WHO), the United Nations Childrens' Fund (UNICEF) and the International Council for Combatting Iodine Deficiency Disorders (ICCIDD) a complete elimination of IDD in Africa is still thought to be a distant dream (see, e.g., Saha et al. 2019). This is so, because of the interplay of an array of geochemobiological variables [e.g., preponderance of fluorine activity over iodine, the effect of *goitrogenic* substances, loss of iodine bioavailability or disturbance in the iodine transport apparatus *whose mechanisms of interaction are not yet fully understood*; thus underscoring the need for further research on the Medical Geology of iodine (Saha et al. 2019; Davies 2024).

Box 1. Elimination of iodine deficiency disorders in Africa: The power of multidisciplinary science

---

In today's rapidly evolving scientific era, the multidisciplinary approach is proving to be a powerful strategy that brings together insights, methodologies, and expertise from diverse fields to address complex issues (Levain 2023). The virtual elimination of IDD, one of the most intractable of hormonal dysfunctions in Africans presents a good example of how leveraging the collective expertise of various disciplines can help solve vexing scientific problems.
Modern medicine had initially prescribed absolute levels of iodine [i.e., 140 μg (μg) of iodine a day (NHS, Undated)]. Contemporary geochemists have now characterised the elements' levels in the various geological landscapes, as well as the forms and quantities of the element that eventually reach the centres of the body where it is required for pertinent metabolic processes (thyroid hormone modulation). We note that several metabolic pathways are in specific locations inside the *mitochondria* [membrane-bound cell organelles (mitochondrion, singular) that generate most of the chemical energy needed to power the cell's biochemical reactions]. Loss of iodine bioavailability, such as a dysfunctional transport mechanism, or when the activity of available iodine is inhibited due to its co-occurrence with fluorine (F), or the presence of *goitrogens* may result in reduced antioxidant and anti-inflammatory capacity, as well as impaired immunity and enhanced risk of IDD. The elucidation of these and other geochemobiological dynamics of iodine uptake and metabolism falls within the compass of the Medical Geologist (among other specialists in a multidisciplinary team) whose further contribution in future iodine research should now be clearly affirmed.

Current Status of Geomedical Research on Iodine in Africa

Efforts aimed at completely eliminating IDD in Africa, have triggered a huge volume of research leading up to the year 2024, mainly looking at the burden of IDD in pregnant women in Africa (see, e.g., Businge et al. 2022; Ngounda et al. (2022);. The emerging problem of excess iodine intakes should, however, also be taken into consideration by policy makers and programme implementers, since this may have adverse health effects greater than those induced by iodine deficient diets (Saha et al. 2019). *More population-based studies are needed to investigate iodine intakes of the different population groups in Africa* (Saha et al. 2019; see Section on: "Hypothyroidism", this Chapter").

Goitrogens

Goitrogens are chemical substances that interfere with thyroid hormone which is secreted by the pituitary gland to promote thyroid tissue growth and regulation of thyroid function. Increased secretion of the *thyroid-stimulating hormone* (TSH) (also called *thyrotropin)* can lead to development of goitre and its sequelae. Consumption of foods that contain *goitrogens,* can therefore exacerbate iodine deficiency (Davies 2010).

As recently as 2024, Baffa et al. showed that goitrogenic food consumption and the sex of the child were determinant factors for the occurrence of iodine deficiency among school children in Ethiopia. Common goitrogens or goitrogenic substances include: the tetrafluoroborate ion $[BF_4]^-$, the thiocyanate ion $[SCN]^-$ originating from consumption of poorly detoxified cassava (*Manihot esculenta* Crantz) (see: Thilly et al. 1992; Oluwole and Oludiran 2013; Muleta and Mohammed 2017) and vegetables belonging to the botanical family Cruciferae or Brassicaceae [e.g., kale (called *sukumawiki* in East Africa), cabbage and broccoli)]. Fluoride exposure has the potential to disrupt thyroid functioning, though adequate iodine intake may mitigate this effect (Malin et al. 2018; Shaik et al. 2019). Deficiencies of iron and/or vitamin A may also be goitrogenic (Hess 2010; Schulhai et al. 2024).

Hypothyroidism

High iodine intake levels can also result in symptoms similar to those resulting from iodine deficiency, such as goitre, elevated levels of TSH, and *hypothyroidism*, a thyroid dysfunction observed worldwide (Taylor et al. 2018, 2024). This is so, because high levels of iodine in susceptible individuals retards thyroid hormone synthesis, thereby increasing TSH stimulation, which can produce goitre. *Thyroxine (T4)*, the key hormone produced in the thyroid gland can also induce hypothyroidism if its levels are either too low; or if too high, *hyperthyroidism* results. The first hypothyroidism diagnoses were made during the late nineteenth century, when doctors observed that surgical removal of the thyroid gland resulted in swelling of the hands, face, feet, and the tissues around the eyes. The syndrome was named *myxedema* and the cause was rightly attributed to the absence of thyroid hormones, which are normally produced by the thyroid gland. Hypothyroidism is usually progressive and irreversible. Studies have also shown that excessive iodine intakes cause *thyroiditis* and *thyroid papillary cancer* (Battistella et al. 2022).

During the last decade, studies on thyroid dysfunction in African populations have largely been centred on hypothyroidism and hyperthyroidism. Of these studies, that of Dave et al. (2015) presented the first South African guideline for the management of hypothyroidism in adults. Later, Abu et al. (2019) reviewed the risks of excess iodine intake in Ghana. Aidoo et al. (2024), working in a tertiary care hospital in Accra, Ghana, found primary hyperthyroidism to be the most commonly diagnosed thyroid dysfunction. Other extraneous conditions associated with thyroid dysfunction in African populations are revealed in the work of Hassan-Kadle et al. (2021) in Somalia, who found papillary thyroid cancer and follicular thyroid cancer to be predominant histologies among thyroid malignancies. *In Mansfield et al. (2022), pointed out the lack of knowledge on the prevalence of dyslipidemia [abnormal levels of lipids (fats) in the blood] in overt hypothyroidism among the ethnically diverse, mostly black South African population.* Maphumulo et al. (2023), researching the paediatric services in tertiary hospitals in Pretoria, South Africa, noted the common association between thyroid dysfunction in children and *Downs Syndrome*, compared with the general population.

Geophagy: An Illustrstive Example of "Exotic" Ingestion

Geophagy (or *geophagia*), the habit of eating clay or earth is very prominent among traditional African societies. Iron deficiency may provide part of the answer to why the practice is still very much alive (see: WHO 2023a, b; Muñoz 2024). Despite the voluminous recent literature on geophagy in Africa (e.g., Nyanza et al. 2014; Lar et al. 2015; Ekosse et al. 2017; Ngole-Jeme et al. 2018; Bonglaisin et al. 2022; Davies 2023; Malepe et al. 2023; Malebatja et al. 2024; and many more recent references), however, the practice remains largely misunderstood to date; but it is thought to have both beneficial and deleterious effects. The great deal of research effort given towards elucidation of the various aspects of the phenomenon, has involved entire research groups (e.g., the IGCP Project 545 on "Geophagia in Southern Africa", and others given in the references in: Ekosse et al. 2017; Davies 2023).

Health Benefits of Geophagy: The Iron Supplementation Paradigm

There are many theories about why people practice geophagy, including its possible roles as nutritional supplement, food detoxifier, diarrhoeal pharmaceutical, in reducing over-salivation, spitting and vomiting during pregnancy, as famine food, for.

psychological, psychiatric and cultural reasons, and so on (Abrahams 2012). Aside from the benefits that soils can impart to the consumer, however, very serious health problems may result (see: Oliver and Brevik 2023).

Of the numerous aetiological theories about geophagy, the "nutrient supplement paradigm" (iron deficiency anaemia) is perhaps the most widely accepted. The RDA of iron for all age groups of men and postmenopausal women is 8 mg/day; the RDA for premenopausal women is 18 mg/day (Institute of Medicine 2001).

Anaemia in pregnancy is a major public health concern, and the World Health Organisation estimates that 37% of pregnancies are affected by anaemia (WHO 2023a, b; Muñoz 2024). During pregnancy, the volume of blood in the body increases, and so does the amount of iron that is needed. As a result of this expansion, and to meet the needs of the foetus and placenta, the amount of iron that women need increases during pregnancy (Sangkhae et al. 2023).

Demerits of Geophagy

The demerits of geophagy are also closely linked to the amount of ingested earth material. These may include helminth infection (especially among children), electrolyte disturbances, tooth damage, blockage of the intestines, constipation, abdominal pains and metal poisoning (Davies 2023). Among pregnant women practitioners, premature birth, neural tube defects and low birth weight have also been observed (see, e.g., Bonglaisin et al. 2022).

Fluorine Deficiency and Excess

The range in intakes producing detrimental or beneficial effects are not far apart (WHO, Undated). The daily tolerable upper intake levels (UL) for fluoride (F^-) for both male and female over 9 years of age is 10 mg (Institute of Medicine, 1997). The World Health Organisation has set a safe limit for fluoride in drinking water of 1.5 mg/L (WHO 2004).

In low concentration (0.5–1.0 mg/L), fluoride is needed by humans for healthy development of bones and teeth. However, at concentrations in excess of 1.5 mg/L several maladies occur. Dental and skeletal fluorosis are the major fluoride-related diseases commonly reported in fluoride-enriched groundwaters in many parts of the world ((Nungula et al. 2025), not least in Africa, where high fluoride contents occur in groundwaters in and around the East African Rift Valley.

The widespread occurrence of undesirable physiological consequences of fluoride exposure in so many African countries, coupled with improvements in analytical techniques for fluoride in biological and other environmental samples (e.g., Srivastava 2025) has triggered a flurry of research on the subject. during the last decade (see, e.g., Onipe et al. 2020; Onipe et al. 2021; Podgorski and Berg 2022; Andreah et al. 2023; Ashong et al. 2024; Kazapoe et al. 2024). Results from these studies have led to the suggestion that food crops can substantially contribute to *fluorosis* in African countries, especially during childhood; *reaffirming the necessity for more studies on fluoride contents in African dietary constituents.*

In Africa, some staple foods such as sorghum, millet and certain vegetables (including kale, spinach, and cowpea leaves) can contain high fluoride levels particularly in regions with naturally occurring fluoride-rich water sources, such as parts of the East African Rift Valley (Gevera et al. 2022; Nelima et al. 2023).

In Africa, fluoride supplementation for dental health is done mainly through community-based methods such as use of fluoridated water and salt. Professionally administered fluoride gels and varnishes are also used, as are self-administered options like the use of toothpaste and oral rinses (see, e.g., Mulder 2018).

Research on defluoridation in Africa in the last decades has centred around the reduction of excessive levels of fluoride in drinking water, mainly in the East African Rift Valley and in North Africa. The major objective has been the prevention of fluorosis. Commonly used methods include the Nalgonda technique, activated alumina and electrocoagulation. Many of these technologies have high initial and operational costs and require skilled operators, making them prohibitive in Africa. This underscores the need for developing applicable and affordable defluoridation technologies for small communities and households. Indeed, during the last decade, newer approaches have appeared and are beginning to gain ground, e.g., the use of harvested and blended rainwater (Marwa et al. 2018), the FLOWERED Defluoridator Device (FDD) (Odini et al. 2020) and the use of carbonised bone as a green adsorbent (Moloi et al. 2024).

The Life Essential Selenium

There is a dearth of research on the circulation of environmental selenium (Se) in Africa, despite evidence of widespread selenium deficiency in the Continents' surface environment, especially on its southern flanks, and the element's probable link with HIV diffusion.

As much as selenium is essential to the health of humans and other animals in trace amounts, it is also harmful in excess. Of all the elements, selenium has one of the narrowest ranges between dietary deficiency (<40 μg/d) and toxicity (>400 μg/ d) and so, careful intake by humans is required to balance the association between health significance and environmental exposure limits (Fordyce, 2007). The RDA for both men and women is given as 55 μg/day (Institute of Medicine 2000).

Realisation of the importance of selenium in nutrition and its deficiency in soils of Sub-Saharan Africa (SSA), especially southern Africa, triggered a surge in research effort in this domain in the last two decades (see, e.g., Hurst et al. 2013). However, research and publications on the possible links between selenium deficiency and HIV/AIDS in southern Africa petered off during the past decade amid controversial evidence on the veracity of this link (see: Kamwesiga et al. 2011; Lukusa et al. 2021; Rollin et al. 2021; Takata et al. (2022); Mutonhodza et al. 2022; Chilala et al. 2024; Oladeji et al. 2024).

The Bill and Melinda Gates Foundation sponsored the GeoNutrition project (2017–2021) (see: Ligowe et al. 2020) the specific aim of which was the mapping of soil-crop-human micronutrient linkages and their uncertainties, including aspects of the environmental circulation of selenium.

Seafood, Brazil nuts, poultry, liver and kidney, as well as other meats provide the richest food sources of selenium (NIH Updated 2024). Grain products and seeds contain selenium, but the amount depends on the selenium content of the soil in which they are grown and the biogeochemical characteristics of the soil.

Nutritional Iron Deficiency and Excess

The human body requires sufficient iron which is crucial for producing red blood cells and transporting oxygen. The daily intake level (DIL) [Recommended Dietary Allowance (RDA)] for iron varies depending on age and sex due to different biological needs and life stages, such as menstruation and pregnancy. Adult men generally need 8 mg/day, while women need 18 mg/day, especially during their reproductive years. Pregnant women require 27 mg/day (NIH 2023).

Iron Deficiency

Iron nutritional deficiency, including *iron deficiency anaemia*, is a significant public health concern in Africa, particularly in sub-Saharan Africa. A substantial portion of the population, especially children and pregnant women, is affected by anaemia, with some countries reporting prevalence rates exceeding 60% in some paediatric populations (e.g., Adane and Getawa 2021; Gebreegziabher and Sidibe 2023). Developmental delays in children and increased maternal and perinatal mortality rates from anaemia have also been reported in recent years (see, e.g., Nampijja et al. 2022; Ringshaw et al. 2025).

Anaemia in pregnancy is a major public health concern, and the World Health Organisation estimates that 37% of pregnancies globally are affected by anaemia (WHO 2023a, b; Muñoz 2024). During pregnancy, the volume of blood in the body increases, and so does the amount of iron that is needed. As a result of this expansion and to meet the needs of the foetus and placenta, the amount of iron that women need increases during pregnancy (Sangkhae et al. 2023).

Diabetes

Iron plays quite a complex role in both the development and progression of diabetes. Widespread iron nutritional deficiency often leads to iron deficiency anaemia whose interplay with diabetes mellitus (DM) and exacerbation of diabetic complications are increasingly recognised as a concern in Africa. This is particularly the case among individuals with type 2 diabetes (Faghir-Ganji et al. 2024).

Iron deficiency can affect glycaemic control in individuals with diabetes while iron overload is a risk factor for type 2 diabetes (T2DM). A multifaceted relationship exists, involving insulin secretion, insulin resistance, and the overall metabolic regulation of glucose and iron.

The prevalence of iron deficiency anaemia is ascribed largely to inadequate African diets, infections, and parasitic infestations (Mwangi et al. 2017). The associated pathogenic features are weakness and impaired growth, motor, and cognitive performance (Kumar et al. 2022).

Treatment regimens for iron deficiency anaemia involve the use of iron supplements or infusions and by addressing any underlying health conditions (Warner and Kamran 2023). Iron-rich foods include both heme iron sources like red meat, poultry, and seafood, and non-heme iron sources like leafy greens, beans, and fortified cereals.

Iron Overload

Iron overload, also known as *haemosiderosis* or *haemochromatosis*, is a condition where the body accumulates excess dietary iron; and is often compounded by a genetic predisposition, potentially leading to organ damage (Porter et al. 2024).

Formerly termed *Bantu siderosis*, *African iron overload* is a type of iron overload disorder prevalent in Southern Africa and Central Africa, which can lead to serious health problems if not adequately managed (Andrews 1999). Diet plays a significant role in managing iron overload, and individuals with this condition should focus on a diet that limits iron intake and absorption.

Africans who drink a traditional beer brewed in non-galvanised steel drums are often the more readily affected (Kew and Asare 2007). It is now thought to actually be two disorders with different causes, possibly compounding each other, viz., dietary iron overload and Bantu siderosis, the genetic side of the disorder (Gordeuk et al. 1999). Another form of iron overload, *gastric siderosis*, can occur due to excessive iron supplementation.

Some African people carry a unique ferroportin (SLC40A1) (a protein that functions as the sole known iron exporter in the body) mutation most notably the Q248H variant that predisposes them to iron overload, making it a kind of *ferroportin disease* (McNamara et al. 2005). This condition (African iron overload) is associated with elevated levels of *serum ferritin*, a protein that stores iron.

In the management of hemochromatosis, Milman (2021) considers that dietary modifications that lower iron intake and decrease iron bioavailability may provide additional measures to reduce iron uptake from the foods and reduce the number of *phlebotomies* [surgical opening or puncture of a vein in order to withdraw blood, to introduce a fluid, or (historically) when letting blood]; but cautions that there is the need for large, prospective, randomised studies that specifically evaluate the effect of dietary interventions.

Podoconiosis: Uptake of High Element Concentrations From Volcanic Clay Soil

Podoconiosis is a non-infectious geochemical disease, characterised by leg swelling (*lymphoedema*). There is evidence of podoconiosis in more than 12 African countries (WHO 2023a, b; Tamiru et al. 2025), with the countries of Ethiopia, Cameroon, Kenya and Rwanda bearing the highest burden.

While the exact causative agent remains unclear some research suggests the exposure to irritant soil, as a risk factor, particularly volcanic red clay (e.g., East African Rift Valley volcanic soils and soils of the Cameroun Volcanic Line) that is rich in certain minerals which occur as microparticles of silica and aluminium silicates, as well as trace elements. The aetiological role for certain elements like zirconium, beryllium, cerium and aluminium, found in the soil of podoconiosis-endemic regions accords with findings in earlier studies, e.g., Frommel et al. (1993); Molla et al. (2014), as well as more recent findings; and continue to suggest the involvement of *high content of trace elements* in the pathogenesis of podoconiosis. These elements penetrate through the skin, driving inflammation and subsequent lymphoedema development in the lower limbs (see: Wanji et al. 2021; Abiso et al. 2023; WHO 2023a, b). Podoconiosis research, particularly that done during the early part of this decade, continues to focus on the role of geochemical factors, more especially, the role of *high amounts* of trace elements in the soil, in the development of this non-filarial lymphoedema (see, e.g., Wanji et al. 2021; Cooper and Nick 2023).

Some recent studies have also examined the interaction between other environmental factors (high altitude, temperature, high seasonal rainfall) and genetic predisposition in the onset of the disease (Wanji et al. 2021; WHO 2023a, b; Negash et al. 2024; WHO 2025). Podoconiosis studies in 2025 and beyond will likely continue to explore the role of *high amounts* of trace elements in the soil, especially in areas with volcanic red clay, in the development of this non-filarial lymphoedema.

Recent evidence suggests that podoconiosis can be managed through public health interventions (Deribe 2018). The key strategies for podoconiosis prevention and control are simple. Primary prevention can be attained through avoidance of contact with irritant soil such as by doning suitable footwear and covering house floors and paving roads; and secondary and tertiary prevention, through lymphoedema morbidity management.

Although much has been learned in recent years, there are still gaps in knowledge of various aspects of podoconiosis studies. Future research will focus on understanding the specific mechanisms by which trace elements, along with other factors like genetic predisposition and soil composition, trigger inflammatory responses leading to the disease (Cooper and Nick 2023). Despite the voluminous literature that now exists on podoconiosis research, significant gaps in knowledge still exist. Aspects such as care for podoconiosis patients and worldwide control require further reliable and relevant future research (Deribe et al. 2020).

The Second International Podoconiosis Conference with the theme of: "From Neglect to Elimination With Country Ownership" took place in Kigali, Rwanda in January this year (2025). [https://ntdsresearch.org/2025/02/11/the-second-podoco niosis-conference-from-neglect-to-elimination/ (Accessed 17.05.2025)]. Rwanda is set to eliminate podoconiosis by 2030 and other countries of the African Continent may follow this lead.

https://rbc.gov.rw/wp/rwanda-set-to-eliminate-podoconiosis-by-2030/ [(Accessed 17.05.2025)].

Conclusions

This Chapter looks at the current status of Medical Geology research around the Continent of Africa in the last decade or so, focussing particularly on nutritional and PTE perturbations in the food chain and disease. The review documents the successes, identifies some knowledge gaps, addresses the pitfalls and examines the current challenges and future perspectives. It is against this backdrop that the design of further research on Medical Geology of the African Continent should be predicated.

Despite the substantial advances that have been made in Medical Geology during the past decade or so, much more needs to be done to maintain the momentum. Bioavailability and bioaccessibility are complex issues that need to be assessed more accurately to evaluate whether or not there may be adverse effects for human health.

There is a need to incorporate physical and chemical parameters in predictive models, which relate toxic metals in soils to human health. However, the scarcity of data providing a direct estimate of the human absorption of PTEs remains a significant limitation on our ability to estimate bioavailability. This is a shortcoming that can, to some extent, be addressed by the use of the most representative animal model. The assessment of bioavailability and bioaccessibility in a monitoring strategy should include the variations in these parameters over time with due cognisance taken of modifications in some soil parameters or land use. Accurate knowledge of these parameters (bioavailability and bioaccessibility) can help in refining risk assessments and offer effective decision-support for managing land where humans are exposed to contaminants.

Exposure dosage and effects are not independent variables, but represent successive phases of a continuum, which links the soil to human health through bioavailability processes including bioaccessibility. *The study of dose response, and developing dose–response models, particularly for populations around major African mining locales is of high importance, for this constitutes the basis for determining "safe", "hazardous" and (where relevant) beneficial levels and dosages for pollutants and foods.*

In the case of iron, as at 2023, a clear causal relationship between dietary heme iron intake and disease risk had not been established (Charlebois and Pantopoulos 2023). *There is therefore the need for further experimental studies to corroborate epidemiological data and explore mechanisms underlying the pathogenicity of high dietary heme iron.*

References

Abiso TL, Abebe Kerbo A, Wolka Woticha E (2023) Epidemiology of podoconiosis in sub-Saharan Africa: a systematic review and meta-analysis. SAGE Open Med 11:205031212311193602. https://doi.org/10.1177/20503121231193602. Accessed 17 May 2025

Abrahams PW (2012) Involuntary soil ingestion and geophagia: a source and sink of mineral nutrients and potentially harmful elements to consumers of earth materials. Appl Geochem 27(5):954–968. https://doi.org/10.1016/j.apgeochem.2011.05.003. Accessed 14 Jan 2025

Abu BAZ, Oldewage-Theron W, Aryeetey RNO (2019) Risks of excess iodine intake in Ghana: Current situation, challenges, and lessons for the future. Ann N Y Acad Sci 1446(1):117–138. https://doi.org/10.1111/nyas.13988. Accessed 04 Feb 2025

Adane T, Getawa S (2021) Anaemia and its associated factors among diabetes mellitus patients in Ethiopia: a systematic review and meta-analysis. Endocrinol Diabet Metab 4:e00260. https://doi.org/10.1002/edm2.260. Accessed 15 Aug 2025

Adhikari S, Struwig M (2024) Concentrations and health risks of selected elements in leafy vegetables: a comparison between roadside open-air markets and large stores in Johannesburg, South Africa. Environ Monit Assess 196:170. https://doi.org/10.1007/s10661-023-12283-6. Accessed 04 Feb 2024

Aidoo ED, Ababio GK, Arko-Boham B, Tagoe EA, Aryee NA (2024) Thyroid dysfunction among patients assessed by thyroid function tests at a tertiary care hospital: a retrospective study. Pan Afric Med J 49:7. https://doi.org/10.11604/pamj.2024.49.7.44173. Accessed 02 Feb 2025

Andrcah K, Tcmbo M, Manda M (2023) Community awareness of dental fluorosis as a health risk associated with fluoride in improved groundwater sources in Mangochi District Malawi. J Water Health 21(2):192–204. https://doi.org/10.2166/wh.2023.210. Accessed 11 Jan 2025

Andrews NC (1999) Disorders of iron metabolism. New England Journal of Medicine 341(26), (1986–1995). 1999. Note Errat New Engl J Med 342:364 only, 2000. https://doi.org/10.1056/NEJM199912233412607. Accessed 6 Aug 2025

Arnot JA, Toose L, Armitage JM, Sangion A, Looky A, Brown TN, Li L, Becker RA (2022) Developing an internal threshold of toxicological concern (iTTC). J Eposure Sci Environ Epidemiol 32(6):877–884. https://doi.org/10.1038/s41370-022-00494-x. Accessed 18 Apr 2024

Ashong GW, Ababio BA, Kwaansa-Ansah EE, Koranteng SK, Muktar GD (2024) Investigation of fluoride concentrations, water quality, and non-carcinogenic health risks of borehole water in bongo district, northern Ghana. Heliyon 10(6):e27554. https://doi.org/10.1016/j.heliyon.2024.e27554. Accessed 11 Jan 2025

Baffa LD, Angaw DA, Abriham ZY, Gashaw M, Agimas MC, Sisay M, Muhammad EA, Mengistu B, Belew AK (2024) Prevalence of iodine deficiency and associated factors among school-age children in Ethiopia: a systematic review and meta-analysis. Syst Rev 13(1):142. https://doi.org/10.1186/s13643-024-02567-4. Accessed 02 Feb 2025

Battistella E, Pomba L, Costantini A, Scapinello A, Toniato A (2022) Hashimoto's thyroiditis and papillary cancer thyroid coexistence exerts a protective effect: a single centre experience. Indian J Surg Oncol 13(1):164–168. https://doi.org/10.1007/s13193-022-01515-9. Accessed 02 Feb 2025

Boahen E, Fosu-Mensah BY, Koranteng SS, Darko DA, Obuobi G, Mensah M (2024) Potentially toxic elements' accumulation and health risk of consuming vegetables cultivated along the Accra-Tema motorway. J Chem https://doi.org/10.1155/2024/6438563. Accessed 04 Feb 2024

Bonglaisin JN, Kunsoan NB, Bonny P, Matchawe C, Tata BN, Nkeunen G, Mbofung CM (2022) Geophagia: Benefits and potential toxicity to human-a review. Front Public Health 10:893831. https://doi.org/10.3389/fpubh.2022.893831. Accessed 11 Jan 2025

Businge CB, Musarurwa HT, Longo-Mbenza B, Kengne AP (2022) The prevalence of insufficient iodine intake in pregnancy in Africa: a systematic review and meta-analysis. Syst Rev 11:231. https://doi.org/10.1186/s13643-022-02072-6. Accessed 13 Jan 2025

Charlebois E, Pantopoulos K (2023) Nutritional aspects of iron in health and disease. Nutrients 15(11):2441. https://doi.org/10.3390/nu15112441. Accessed 15 Aug 2025

Chilala P, Skalickova S, Horky P (2024) Selenium Status of Southern Africa. Nutrients 16(7):975. https://doi.org/10.3390/nu16070975. Accessed 13 Jan 2025

Cooper JN, Nick KE (2023) A geochemical and mineralogical characterization of soils associated with podoconiosis. Environ Geochem Health 45:7791–7812. https://doi.org/10.1007/s10653-023-01625-5. Accessed 17 May 2025

Dave JA, Klisiewicz A, Bayat Z, Mohamed NA, Stevens Z, Kinvig T, Mollentze WF (2015) SEMDSA/ACE-SA guideline for the management of hypothyroidism in adults South African. Fam Pract 57(6):a4399. https://doi.org/10.4102/safp.v57i6.4399. Accessed 02 Feb 2025

Davies TC (1999) The East and Southern Africa regional workshop on geomedicine. Episodes 22(4):303–304. file:///C:/Users/davies.theophilus/Downloads/IUGS022–04–09%20(4).pdf. Accessed 04 Sept 2023

Davies TC (2010) Medical geology in Africa. In: Selinus O, Finkelman RB, Centeno J (eds.) Medical geology–a regional synthesis; Chapter 8, 1st edn. Springer Verlag, Amsterdam, The Netherlands. ISBN-10: 9048134293. pp 199–219

Davies TC (2022) The position of geochemical variables as causal co-factors of diseases of unknown aetiology. Springer Nature Appl Sci 4:236. https://doi.org/10.1007/s42452-022-051 13-w. Accessed 27 Jun 2022

Davies TC (2023) Current status of research and gaps in knowledge of geophagic practices in Africa. Front Nutr 9:1084589. https://www.ncbi.nlm.nih.gov/pmc/articles/PMC9987423/. Accessed 27 Feb 2024

Davies TC (2024) Medical geology of iodine. In: Davies TC (ed) Medical geology of Africa: a research primer, Chapter 10. ISBN: 9780128187487. Medical Geology of Africa: a research primer–1st edn. Elsevier, pp 481–535. https://shop.elsevier.com/books/medical-geology-of-afr ica-a-research-primer/davies/978-0-12-818748-7. (Accessed 28.04.2024)

Deribe K, Mackenzie CD, Newport MJ, Argaw D, Molyneux DH, Davey G (2020) Podoconiosis: key priorities for research and implementation. Trans R Soc Trop Med Hyg 114(12):889–895. https://doi.org/10.1093/trstmh/traa094

Deribe K (2018) Podoconiosis today: challenges and opportunities. Trans R Soc Trop Med Hyg 112(11):473–475

Dictionary of Toxicology (2024) U-shaped dose-response curve. In: Dictionary of toxicology. Springer, Singapore. https://doi.org/10.1007/978-981-99-9283-6_2772. Accessed 08 Feb 2022

Ekosse G-IE, Ngole-Jeme VM, Diko M (2017) Environmental geochemistry of geophagic materials from Free State Province in South Africa. Open Geosci 9(1):114–125. https://doi.org/10.1515/ geo-2017-0009. Accessed 11 Jan 2025

Faghir-Ganji M, Abdolmohammadi N, Nikbina M, Amanollahi A, Ansari-Moghaddam A, Khezri R, Baradaran H (2024) Prevalence of anaemia in patients with diabetes mellitus: a systematic review and meta-analysis. Biomed Environ Sci 37(1):96–107. https://doi.org/10.3967/bes202 4.008

Farinde A (2023) Dose-response relationships. Merck Manual. https://www.merckmanuals.com/ en-ca/professional/clinical-pharmacology/pharmacodynamics/dose-response-relationships. Accessed 08 Feb 2025

Fayiga AO, Ipinmoroti MO, Chirenje T (2018) Environmental pollution in Africa. Environ Dev Sustain 20:41–73. https://doi.org/10.1007/s10668-016-9894-4. Accessed 04 Feb 2024

Fordyce FM (2007) Selenium geochemistry and health. Ambio 36(1):94–97. https://doi.org/10. 2307/4315793

Gaothobogwe K, Ultra V, Rantong G, Majoni S, Manyiwa S, Mokgosi S, Sefatlhi K (2025) Mitigation of potentially toxic elements in corn (Zea Mays) grown in farmlands near Cu-Ni mine in Central Botswana. Soil Sediment Contamin (Formerly J Soil Contamin) 1–27. https://doi.org/ 10.1080/15320383.2025.2454513. Accessed 13 Feb 2025

Gebreegziabher T, Sidibe S (2023) Prevalence and contributing factors of anaemia among children aged 6–24 months and 25–59 months in Mali. J Nutr Sci 12:e112. https://doi.org/10.1017/jns. 2023.93. Accessed 15 Aug 2025

Gevera PK, Cave M, Dowling K, Gikuma-Njuru P, Mouri H (2022) Potential fluoride exposure from selected food crops grown in high fluoride soils in the Makueni County, south-eastern Kenya. Environ Geochem Health 44(12):4703–4717. https://doi.org/10.1007/s10653-022-012 40-w. Accessed 18 Aug 2023

Gordeuk V, Mukiibi J, Hasstedt SJ, Samowitz W, Edwards CQ, West G, et al (1992) Iron overload in Africa. Interaction between a gene and dietary iron content. New England J Med 326(2):95–100. https://doi.org/10.1056/NEJM199201093260204

Grandjean P (2016) Paracelsus revisited: the dose concept in a complex world. Basic Clin Pharmacol Toxicol 119(2):126–132. https://doi.org/10.1111/bcpt.12622. Accessed 01 Feb 2025

Hassan-Kadle MA, Adani AA, Eker HH, Keles E, Muse Osman M, Mahdi Ahmed H, Görçin Karaketir Ş (2021) Spectrum and prevalence of thyroid diseases at a tertiary referral hospital in

Mogadishu, Somalia: a retrospective study of 976 cases. Int J Endocrinol 2021:7154250. https://doi.org/10.1155/2021/7154250. Accessed 02 Feb 2025

Hess SY (2010) The impact of common micronutrient deficiencies on iodine and thyroid metabolism: the evidence from human studies. Best Pract Res Clin Endocrinol Metab 24(1):117–132. https://doi.org/10.1016/j.beem.2009.08.012. Accessed 02 Feb 2025

Hoseinzadeh E, Taha P (2024) Environmental iodine as a natural iodine intake in humans and environmental pollution index: a scientometric and updated mini review. Int J Environ Health Res 34(10):3600–3614. https://doi.org/10.1080/09603123.2024.2312546. Accessed 28 Apr 2024

Hurst R, Siyame EW, Young SD, Chilimba AD, Joy EJ, Black CR, Ander EL, Watts MJ, Chilima B, Gondwe J, Kang'ombe D, Stein AJ, Fairweather-Tait SJ, Gibson RS, Kalimbira AA, Broadley MR (2013) Soil-type influences human selenium status and underlies widespread selenium deficiency risks in Malawi. Sci Rep 3:1425. https://doi.org/10.1038/srep01425. Accessed 08 Feb 2022

Ibia TO, Chude VO, Nafiu A, Essien GN (2020) Importance of micronutrients in agriculture in Africa and their impacts on health: a review. J Agric Environ 16(1):155–172. https://www.ajol.info/index.php/jagrenv/article/view/235035/222046. Accessed 09 Jan 2025

Idini A, Frau F, Gutierrez L, Dore E, Nocella G, Ghiglieri G (2020) Application of octacalcium phosphate with an innovative household-scale defluoridator prototype and behavioral determinants of its adoption in rural communities of the East African Rift Valley. Integr Environ Assess Manag 16(6):856–870. https://doi.org/10.1002/ieam.4262. Accessed 18 Aug 2025

Institute of Medicine (US) (Food and Nutrition Board) (2001) Dietary reference intakes for vitamin a, vitamin k, arsenic, boron, chromium, copper, iodine, iron, manganese, molybdenum, nickel, silicon, vanadium, and zinc. Report of the panel on micronutrients. National Academy Press (US), Washington DC. https://doi.org/10.17226/10026

Institute of Medicine (US) (Standing Committee on the Scientific Evaluation of Dietary Reference Intakes) (1997) Dietary reference intakes for calcium, phosphorus, magnesium, vitamin D, and fluoride. National Academies Press (US). https://doi.org/10.17226/5776

Institute of Medicine (US) Panel on Dietary Antioxidants and Related Compounds (2000) Dietary reference intakes for Vitamin C, Vitamin E, selenium, and carotenoids. National Academies Press (US), Selenium, Washington DC, p 7. https://www.ncbi.nlm.nih.gov/books/NBK225470/. Accessed 13 Jan 2025

Islam MR, Akash S, Jony MH, Alam MN, Nowrin FT, Rahman MM, Rauf A, Thiruvengadam M (2023) Exploring the potential function of trace elements in human health: a therapeutic perspective. Mol Cell Biochem 478(10):2141–2171. https://doi.org/10.1007/s11010-022-046 38-3. Accessed 8 Feb 2022

John SOO, Olukotun SF, Kupi TG, Mathuthu M (2024) Health risk assessment of heavy metals and physicochemical parameters in natural mineral bottled drinking water using ICP-MS in South Africa. Appl Water Sci 14:202. https://doi.org/10.1007/s13201-024-02267-3. Accessed 04 Feb 2024

Jomova K, Makova M, Alomar SY, Alwasel SH, Nepovimova E, Kuca K, Rhodes CJ, Valko M (2022) Essential metals in health and disease. Chem Biol Interact 367:110173. https://doi.org/10.1016/j.cbi.2022.110173. Accessed 08 Feb 2022

Kassim IA, Ruth LJ, Creeke PI, Gnat D, Abdalla F, Seal AJ (2012) Excessive iodine intake during pregnancy in Somali refugees. Matern Child Nutr 8(1):49–56. https://doi.org/10.1111/j.1740-8709.2010.00259.x. Accessed 14 Aug 2025

Kazapoe RW, Yahans Amuah EE, Dankwa P, Fynn OF, Addai MO, Berdie BS, Douti NB (2024) Fluoride in groundwater sources in Ghana: A multifaceted and country-wide review. Heliyon 10(13):e33744. https://doi.org/10.1016/j.heliyon.2024.e33744. Accessed 11 Jan 2025

Kew MC, Asare GA (2007) Dietary iron overload in the African and hepatocellular carcinoma. Liver Int Offic J Int Assoc Study Liver 27(6):735–741. https://doi.org/10.1111/j.1478-3231.2007.015 15.x. Accessed 15 Aug 2025

Kihara J, Bolo P, Kinyua M, Rurinda J, Piikki K (2020) Micronutrient deficiencies in African soils and the human nutritional nexus: opportunities with staple crops. Environ Geochem Health 42(9):3015–3033. https://doi.org/10.1007/s10653-019-00499-w. Accessed 09 Jan 2025

Kumar SB, Arnipalli SR, Mehta P, Carrau S, Ziouzenkova O (2022) Iron deficiency anemia: efficacy and limitations of nutritional and comprehensive mitigation strategies. Nutrients 14(14):2976. https://doi.org/10.3390/nu14142976. Accessed 15 Aug 2025

Lar UA, Agene JI, Umar AI (2015) Geophagic clay materials from Nigeria: a potential source of heavy metals and human health implications in mostly women and children who practice it. Environ Geochem Health 37(2):363–375. https://doi.org/10.1007/s10653-014-9653-0. Accessed 11 Jan 2025

Levain R (2023) Embracing the power of the multidisciplinary approach breaking boundaries and fostering innovation. Short Commun–JBR J Interdiscipl Med Dent Sci 6:3. https://www.openaccessjournals.com/articles/embracing-the-power-of-the-multidisciplinary-approach-breaking-boundaries-and-fostering-innovation-16444.html. Accessed 04 Feb 2025

Ligowe IS, Phiri FP, Ander EL, Bailey EH, Chilimba A, Gashu D, Joy E, Lark RM et al (2020) Selenium deficiency risks in sub-Saharan African food systems and their geospatial linkages. Proc Nutr Soc 79(4):457–467. https://doi.org/10.1017/S0029665120006904

Lukusa K, Abubeker H, Lehloenya KC (2021) Dietary selenium supplementation, clarified egg yolk extender and slow cooling improve cryopreserved sperm characteristics of Saanen buck. Asian Pacific J Reprod 10(1):43–48. https://doi.org/10.4103/2305-0500.306437

Malebatja MF, Randa MB, Mokgatle MM, Oguntibeju OO (2024) Nurses' perspectives of geophagic women of childbearing age accessing healthcare in the reproductive healthcare services in Tshwane District Gauteng Province: an exploratory study. Women 4:541–551. https://doi.org/10.3390/women4040040. Accessed 11 Jan 2025

Malepe RE, Candeias C, Mouri H (2023) Geophagy and its potential human health implications–a review of some cases from South Africa. J Afr Earth Sc 200:104848. https://doi.org/10.1016/j.jafrearsci.2023.104848. Accessed 19 May 2023

Malin AJ, Riddell J, McCague H, Till C (2018) Fluoride exposure and thyroid function among adults living in Canada: effect modification by iodine status. Environ Int 121(Pt 1):667–674. https://doi.org/10.1016/j.envint.2018.09.026. Accessed 02 Feb 2025

Mansfield BS, Bhana S, Raal FJ (2022) Dyslipidemia in South African patients with hypothyroidism. J Clin Transl Endocrinol 29:100302. https://doi.org/10.1016/j.jcte.2022.100302. Accessed 02 Feb 2025

Maphumulo SF, Honey EM, Abdelatif N, Karsas M (2023) The prevalence and spectrum of thyroid dysfunction among children with Down syndrome attending the paediatric services at two tertiary hospitals in Pretoria, South Africa. South African J Child Health 17(4):e2007. https://doi.org/10.7196/SAJCH.2023.v17i4.2007. Accessed 02 Feb 2025

Marwa J, Lufingo M, Noubactep C, Machunda R (2018) Defeating fluorosis in the East African Rift Valley: transforming the Kilimanjaro into a Rainwater Harvesting Park. Sustainability 10:4194. https://doi.org/10.3390/su10114194. Accessed 18 Aug 2025

McNamara L, Gordeuk VR, MacPhail AP (2005) Ferroportin (Q248H) mutations in African families with dietary iron overload. J Gastroenterol Hepatol 20(12):1855–1858. https://doi.org/10.1111/j.1440-1746.2005.03930.x. Accessed 15 Aug 2025

Milman NT (2021) Managing genetic hemochromatosis: an overview of dietary measures, which may reduce intestinal iron absorption in persons with iron overload. Gastroenterol Res 14(2):66–80. https://doi.org/10.14740/gr1366. Accessed 15 Aug 2025

Moffett DB, Mumtaz MM, Sullivan Jr DW, Whittaker MH (2022) General considerations of dose-effect and dose-response relationships. In: Nordberg GF, Costa M (eds.) Handbook on the toxicology of metals, 5th edn, Volume I: General considerations, Chapter 13, pp 299–317. https://doi.org/10.1016/B978-0-12-823292-7.00019-X. Accessed 02 Feb 2025

Molla YB, Wardrop NA, Le Blond JS, Baxter P, Newport MJ, Atkinson PM, Davey G (2014) Modelling environmental factors correlated with podoconiosis: a geospatial study of non-filarial elephantiasis. Int J Health Geograph 13:24. https://doi.org/10.1186/1476-072X-13-24. Accessed 17 May 2025

Moloi S, Okonkwo OK, Jansen R (2024) Defluoridation of water through the application of carbonised bone as a green adsorbent: a review. South African J Sci 120(1/2). https://doi.org/10.17159/sajs.2024/12879. Accessed 18 Aug 2025

Mulder R (2018) Systemic fluoride supplementation in South Africa–Updated guidelines for practitioners. Int Dent–Afric Edn 8(6). https://www.moderndentistrymedia.com/dec_jan2019/mulder.pdf. Accessed 18 Aug 2023

Muleta F, Mohammed (2017) Determination of cyanide concentration levels in different cassava varieties in selected iodine deficiency disordered (IDD) areas of Wolaita Zone, Southern Ethiopia. J Nat Sci Res 7(3). ISSN 2224–3186 (Paper) ISSN 2225–0921 (Online)

Muñoz JL (2024) Anaemia in pregnancy. MSD manual, professional version. https://www.msdmanuals.com/professional/gynecology-and-obstetrics/approach-to-the-pregnant-woman-and-prenatal-care/anemia-in-pregnancy. Accessed 19 May 2025

Mutonhodza B, Joy EJM, Bailey EH, Lark MR, Kangara MGM, Broadley MR, Matsungo TM, Chopera P (2022) Linkages between soil, crop, livestock, and human selenium status in Sub-Saharan Africa: a scoping review. Int J Food Sci Technol 57:6336–6349. https://doi.org/10.1111/ijfs.15979. Accessed 13 Jan 2025

Mwangi MN, Phiri KS, Abkari A, Gbané M, Bourdet-Sicard R, Braesco VA, Zimmermann MB, Prentice AM (2017) Iron for Africa-report of an expert workshop. Nutrients 9(6):576. https://doi.org/10.3390/nu9060576. Accessed 15 Aug 2025

Nampijja M, Mutua AM, Elliott AM, Muriuki JM, Abubakar A, Webb EL, Atkinson SH (2022) Low haemoglobin levels are associated with reduced psychomotor and language abilities in young Ugandan children. Nutrients 14(7):1452. https://doi.org/10.3390/nu14071452. Accessed 16 Aug 2025

Nana AS, Falkenberg T, Rechenburg A, Ntajal J, Kamau JW, Ayo A, Borgemeister C (2023) Seasonal variation and risks of potentially toxic elements in agricultural lowlands of central Cameroon. Environ Geochem Health 45:4007–4023. https://doi.org/10.1007/s10653-022-01473-9. Accessed 04 Feb 2025

NHS (National Health Service) (Undated). Iodine: Vitamins and Minerals. https://www.nhs.uk/conditions/vitamins-and-minerals/iodine/. (Accessed 13.01.2025)

National Institutes of Health (NIH) (2024) Selenium: fact sheet for health professionals. https://ods.od.nih.gov/factsheets/SeleniumHealthProfessional/#:~:text=Adequate%20Intake%20(AI)-,Sources%20of%20Selenium,%2C%20and%20eggs%20%5B14%5D. Accessed 16 Aug 2025

National Institutes of Health, Office of Dietary Supplements (NIH) (2023) Iron: fact sheet for consumers. https://ods.od.nih.gov/factsheets/Iron-Consumer/. Accessed 16 Aug 2025

Negash M, Chanyalew M, Girma T, Alemu F, Alcantara D, Towler B, Davey G, Boyton RJ, Altmann DM, Howe R, Newport MJ (2024) Evidence for immune activation in pathogenesis of the HLA class II associated disease, podoconiosis. Nat Commun 15(1):2020. https://doi.org/10.1038/s41467-024-46347-z. Accessed 17 May 2025

Nelima D, Wambu EW, Kituyi JL (2023) Fluoride distribution in selected foodstuffs from Nakuru County, Kenya, and the risk factors for its human overexposure. Sci Rep 13:15295. https://doi.org/10.1038/s41598-023-41601-8. Accessed 18 Aug 2023

Newell J, Cox SF, Doherty R (2025) Oral bioaccessibility trends for As, Cd, Cr, Ni, and Pb in vegetables grown in contaminated soils: a systematic review. Urban Agric Reg Food Syst 10(1):e70009. https://doi.org/10.1002/uar2.70009. Accessed 03 Feb 2025

Ngole-Jeme VM, Ekosse G-IE, Songca S (2018) An analysis of human exposure to trace elements from deliberate soil ingestion and associated health risks. J Eposure Sci Environ Epidemiol 28:55–63. https://doi.org/10.1038/jes.2016.67. Accessed 11 Jan 2025

Ngounda J, Baumgartner J, Nel M, Walsh CM (2022) Iodine status of pregnant women residing in the urban Free State Province of South Africa is borderline adequate: the NuEMI study. Nutr

Res (New York, N.Y.) 98:18–26. https://doi.org/10.1016/j.nutres.2021.12.002. Accessed 13 Jan 2025

Nungula EZ, Ngaiza VV, Chappa LR, Mwadalu R, Nyambele KA, Uddin S, Rajan S, Sow S, et al (2025) Fluoride in groundwater: Causes, implications and mitigation measures. In: Sharma K (ed) Fluorides in drinking water. Environmental science and engineering. Springer, Cham. https://doi.org/10.1007/978-3-031-77247-4_11. Accessed 08 Feb 2025

Nyanza EC, Joseph M, Premji SS, Thomas DS, Mannion C (2014) Geophagy practices and the content of chemical elements in the soil eaten by pregnant women in artisanal and small scale gold mining communities in Tanzania. BMC Pregn Childbirth 14:144. https://doi.org/10.1186/1471-2393-14-144. Accessed 11 Jan 2025

Oladeji OM, Magoro K, Mugivhisa LL, Olowoyo JO (2024) Selenium and other heavy metal levels in different rice brands commonly consumed in Pretoria South Africa. Heliyon 10(9):e29757. https://doi.org/10.1016/j.heliyon.2024.e29757. Accessed 13 Jan 2025

Oliver MA, Brevik EC (2023) Soil and human health. In Encyclopedia of soils in the environment. Elsevier, pp 555–571. https://doi.org/10.1016/B978-0-12-822974-3.00066-5

Oluwole OS, Oludiran AO (2013) Normative concentrations of urine thiocyanate in cassava eating communities in Nigeria. Int J Food Sci Nutr 64(8):1036–1041. https://doi.org/10.3109/09637486.2013.825697. Accessed 02 Feb 2025

Onipe T, Edokpayi JN, Odiyo JO (2020) A review on the potential sources and health implications of fluoride in groundwater of Sub-Saharan Africa. J Environ Sci Health Part A 55(9):1078–1093. https://doi.org/10.1080/10934529.2020.1770516

Onipe T, Edokpayi JN, Odiyo JO (2021) Geochemical characterization and assessment of fluoride sources in groundwater of Siloam area, Limpopo Province, South Africa. Sci Rep 11:14000. https://doi.org/10.1038/s41598-021-93385-4. Accessed 30 Oct 2023

Panzeri C, Pecoraro L, Dianin A, Sboarina A, Arnone OC, Piacentini G, Pietrobelli A (2024) Potential micronutrient deficiencies in the first 1000 days of life: the pediatrician on the side of the weakest. Curr Obes Rep https://doi.org/10.1007/s13679-024-00554-3. Accessed 18 Apr 2024

Peijnenburg WJ, Jager T (2003) Monitoring approaches to assess bioaccessibility and bioavailability of metals: matrix issues. Ecotoxicol Environ Saf 56(1):63–77. https://doi.org/10.1016/s0147-6513(03)00051-4. Accessed 17 Apr 2024

Petruzzelli G, Pedron F, Rosellini I (2020) Bioavailability and bioaccessibility in soil: a short Petruzzelli review and a case study. AIMS Environ Sci 208–225. https://doi.org/10.3934/environsci.2020013

Podgorski J, Berg M (2022) Global analysis and prediction of fluoride in groundwater. Nat Commun 13:4232. https://doi.org/10.1038/s41467-022-31940-x. Accessed 11 Jan 2025

Porter JL, Rawla P (2024) Hemochromatosis. In: StatPearls [Internet]. StatPearls Publishing, Treasure Island (FL). https://www.ncbi.nlm.nih.gov/books/NBK430862/. Accessed 06 Aug 2025

Prashanth L, Kattapagari KK, Teja CR, Baddam VRR, Prasad LK (2015) A review on role of essential trace elements in health and disease. J Dr. NTR Univ Health Sci 4(2):75–85. https://doi.org/10.4103/2277-8632.158577

Price G, Patel DA (2023) Drug bioavailability. In: StatPearls. StatPearls Publishing. https://www.ncbi.nlm.nih.gov/books/NBK557852/#article-18289.s2. Accessed 17 Apr 2024

Ringshaw JE, Hendrikse CJ, Wedderburn, Bradford LE, Williams SR, Nyakonda CN, et al (2025) Persistent impact of antenatal maternal anaemia on child brain structure at 6–7 years of age: a South African child health study. BMC Med 23:94. https://doi.org/10.1186/s12916-024-03838-6. Accessed 16 Aug 2025

Ripanda AS, Hossein M, Rwiza MJ, Nyanza EC, Selemani JR, Nkrumah S, Bakari R, Alfred MS, Machunda RL, Vuai SA (2025) Combatting toxic chemical elements pollution for sub-Saharan Africa's ecological health. Environ Pollut Manag 2:42–62. https://doi.org/10.1016/j.epm.2025.01.003. Accessed 13 Feb 2025

Röllin HB, Channa K, Olutola B, Odland JØ (2021) Selenium status, its interaction with selected essential and toxic elements, and a possible sex-dependent response in utero, in a South African birth cohort. Int J Environ Res Public Health 18(16):8344. https://doi.org/10.3390/ijerph181 68344. Accessed 14 Jan 2025

Saha S, Abu BAZ, Jamshidi-Naeini Y, Mukherjee U, Miller M, Peng LL, Oldewage-Theron W (2019) Is iodine deficiency still a problem in sub-Saharan Africa?: a review. Proc Nutr Soc 78(4):554–566. https://doi.org/10.1017/S0029665118002859. Accessed 13 Jan 2025

Sangkhae V, Fisher AL, Ganz T, Nemeth E (2023) Iron homeostasis during pregnancy: Maternal, placental, and foetal regulatory mechanisms. Annu Rev Nutr 43:279–300. https://doi.org/10.1146/annurev-nutr-061021-030404. Accessed 16 Aug 2025

Schulhai A-M, Rotondo R, Petraroli M, Patianna V, Predieri B, Iughetti L, Esposito S, Street ME (2024) The role of nutrition on thyroid function. Nutrients 16:2496. https://doi.org/10.3390/nu1 6152496. Accessed 02 Feb 2025

Shaik N, Shanbhog R, Nandlal B, Tippeswamy HM (2019) Fluoride and thyroid function in children resident of naturally fluoridated areas consuming different levels of fluoride in drinking water: an observational study. Contemp Clin Dent 10(1):24–30. https://doi.org/10.4103/ccd.ccd_1. Accessed 02 Feb 2025

Sharma K (ed.) (2025) Fluorides in drinking water. Springer Nature Switzerland, pp 391–442. https://doi.org/10.1007/978-3-031-77247-4_16. Accessed 08 Feb 2025

Siddig MMS, Brevik EC, Sauer D (2025) Human health risk assessment from potentially toxic elements in the soils of Sudan: a meta-analysis. Sci Total Environ 958:178196. https://doi.org/10.1016/j.scitotenv.2024.178196. Accessed 04 Feb 2024

Srivastava N (2025) Role of recent advanced biological technology in removal of fluoride. Part Book Ser Environ Sci Eng Environ Sci Eng Part F4036:391–442

Takata N, Myburgh J, Botha A, Nomngongo PN (2022) The importance and status of the micronutrient selenium in South Africa: a review. Environ Geochem Health 44(11):3703–3723. https://doi.org/10.1007/s10653-021-01126-3. Accessed 17 14 2025

Tamiru S, Lulu Y, Bidira K, Amsalu B, Duguma A, Nigusu Y, Gezimu W (2025) Healthcare professionals' knowledge, attitudes, and practice of podoconiosis management and associated factors in public hospitals in Ilu Ababor and Buno Bedelle zones, Southwest Ethiopia: a cross-sectional study. Front Public Health 13:1454979. https://doi.org/10.3389/fpubh.2025.1454979. Accessed 17 May 2025

Taylor PN, Albrecht D, Scholz A, Gutierrez-Buey G, Lazarus JH, Dayan CM, Okosieme OE (2018) Global epidemiology of hyperthyroidism and hypothyroidism. Nat Rev Endocrinol 14(5):301–316. https://doi.org/10.1038/nrendo.2018.18. Accessed 02 Feb 2025

Taylor PN, Medici MM, Hubalewska-Dydejczyk A, Boelaert K (2024) Hypothyroidism. Lancet (London, England) 404(10460):1347–1364. https://doi.org/10.1016/S0140-6736(24)01614-3. Accessed 02 Feb 2025

Thilly CH, Vanderpas JB, Bebe N, Ntambue K, Contempre B, Swennen B, Moreno-Reyes R, Bourdoux P, Delange F (1992) Iodine deficiency, other trace elements, and goitrogenic factors in the etiopathogeny of iodine deficiency disorders (IDD). Biol Trace Elem Res 32:229–243. https://doi.org/10.1007/BF02784606. Accessed 02 Feb 2025

Tsatsakis AM, Vassilopoulou L, Kovatsi L, Tsitsimpikou C, Karamanou M, Leon G, Liesivuori J, Hayes AW, Spandidos DA (2018) The dose response principle from philosophy to modern toxicology: the impact of ancient philosophy and medicine in modern toxicology science. Toxicol Rep 5:1107–1113. https://doi.org/10.1016/j.toxrep.2018.10.001. Accessed 18 Apr 2024

Wanji S, Deribe K, Minich J, Debrah AY, Kalinga A, Kroidl I, Luguet A, Hoerauf A, Ritter M (2021) Podoconiosis-from known to unknown: obstacles to tackle. Acta Trop 219:105918. https://doi.org/10.1016/j.actatropica.2021.105918. Accessed 17 May 2025

Warner MJ, Kamran MT (2023) Iron deficiency anaemia. [Updated 2023]. In: StatPearls [Internet]. StatPearls Publishing, Treasure Island (FL). https://www.ncbi.nlm.nih.gov/books/NBK448065/. Accessed 16 Aug 2025

World Health Organisation (WHO) (2004) Fluoride in drinking-water: background document for development of WHO guidelines for drinking-water quality. WHO/SDE/WSH/03.04/96. https://www.who.int/docs/default-source/wash-documents/wash-chemicals/fluoride-background-document.pdf. Accessed 11 Jan 2025

World Health Organisation (WHO) (2023a) Fact sheet, anaemia. https://www.who.int/news-room/fact-sheets/detail/anaemia. Accessed 08 Jan 2025

World Health Organisation (WHO) (2023b) Podoconiosis (non-filarial lymphoedema). WHO, Geneva. https://www.who.int/news-room/fact-sheets/detail/podoconiosis-(non-filarial-lymphoedema). Accessed 17 May 2025

World Health Organisation (WHO) (2025) Control of neglected tropical diseases: podoconiosis: endemic non-filarial elephantiasis. WHO, Geneva. https://www.who.int/teams/control-of-neglected-tropical-diseases/lymphatic. Accessed 17 May 2025

Zimmermann MB (2020) Iodine and the iodine deficiency disorders. In: Present knowledge in nutrition, vol 1, 11th edn. Basic nutrition and metabolism, Chapter 25. Elsevier, Amsterdam, The Netherlands, pp 429–441. https://doi.org/10.1016/B978-0-323-66162-1.00025-1. Accessed 14 Aug 2025

Chapter 4
Mercury Emission in Small-Scale Gold Panning and Health Effects: South Africa Case Description

A. M. Msomi and T. C. Davies

Abstract Given its well-established impacts on human health, its inherent toxicity, as well as its tendency to accumulate in living organisms, mercury (Hg) pollution is of grave concern to humanity. Notable known toxicity of this non-essential element includes neurological and carcinogenic effects. The elimination of mercury in humans is therefore of considerable importance worldwide. The upsurge in independent artisanal and small-scale gold mining (ASGM) activities in South Africa is contributing towards the provision of employment and generation of wealth. However, economic gains are countermanded by health degradation and destruction of vital industries such as farming and fishing. Significant quantities of mercury are being used for amalgamation of gold (Au) (constituting a major source of mercury emission into the environment), despite the continued introduction of more efficient mining techniques. Our systematic review based on the PRISMA Protocol and backed by documented medical registry data, does show that occupational mercury exposure from ASGM activities in South Africa still (as at 2025) produces signs and symptoms related to neuro-psychological disorders, ataxia, tremor or memory problems. The absence of environmental concern or the lack of enforcement of regulations also persists in many instances. Geochemical surveys prior to operation and close monitoring to quantify emissions during various stages of the gold mining process can minimise contamination. Remediation schemes can also be devised and monitored by the application of sound geochemical techniques for identifying primary sources of mercury release, its mechanism of transport and other toxicokinetic variables. Strengthening efforts by the South African Government at formalising this industry will ensure rapid realisation of these proposals.

Keywords Mercury emissions · Artisanal and small-scale gold mining · Combatting health effects · Site remediation · South Africa

A. M. Msomi · T. C. Davies (✉)
Faculty of Applied and Health Sciences, Mangosuthu University of Technology, Umlazi, KwaZulu Natal Province, Republic of South Africa
e-mail: daviestheophilus2025@yahoo.com

© The Author(s), under exclusive license to Springer Nature Switzerland AG 2026
T. C. Davies (ed.), *Recent Advances in Medical Geology Research in Africa: A Decadal View*, SpringerBriefs in Earth System Sciences,
https://doi.org/10.1007/978-3-032-13754-8_4

Introduction

During the negotiations of the Minamata Convention, the United Nations Environment Programme (UNEP) noted that South Africa is a major emitter of mercury (Hg) on the African continent (The United Nations Environment Programme [UNEP] 2013). Quite appositely, as concern over mercury (Hg) toxicity continues to grow worldwide, South Africa has been at the forefront of global recognition of mercury as a severe environmental pollutant, and ways to combat its toxicity. These efforts culminated last year (2024) in the staging of the "16th International Conference on Mercury as a Global Pollutant (ICMGP)" in Cape Town, South Africa, from 21 to 26 July 2024 [https://minamataconvention.org/sites/default/files/documents/2024-05/MO240502_ICMGP.pdf (Accessed 26.10.2024)]. The programme of the 16th ICMGP included a set of synthesis papers on international policy on controlling mercury emissions and the importance of mercury science in implementing the Minamata Convention.

The current usage of mercury in artisanal and small-scale gold mining (ASGM) in South Africa remains significant, with substantial quantities of mercury being utilised for each ton of gold (Au) processed, despite the introduction of more efficient mining techniques. In line with the global trend, the occupational and environmental hazards associated with the informal ASGM trade in the Country remain largely undocumented (Street 2015; Ledwaba 2017; Bester 2023). *Patterns of disease, injury and premature death in artisanal and small-scale gold miners are still not clearly defined* (Landrigan et al. 2022).

The colloquial South African name for "artisanal" gold miners is "Zama-Zama" a nomenclature that the South African Government associates with illegality (Nhlengetwa and Hein 2015); but Makhetha (2023) prefers the name "unlicensed miners" or "unregulated miners". Zama Zamas access disused gold mines, to harness the small quantities of gold ore left over by large scale mining companies.

The health effects of mercury exposure in small-scale gold mining in South Africa include neuro-psychological disorders such as ataxia, tremors and memory problems (see, e.g., George et al. 2023). Additionally, other reported symptoms such as hair loss or pain remain largely unspecific. These health issues are known to persist even today, because miners still use significant quantities of mercury in the gold extraction process. It is essential to address these health effects through measures such as geochemical surveys, and close monitoring of emissions and disease surveillance programmes, to enable the design of exposure limiting measures and the implementation of remediation schemes. Strengthening of efforts by the South African Government to formalise the industry is also crucial to ensure the rapid realisation of these proposals (Bester 2023).

Methodology

In this study, we systematically reviewed articles that reported new techniques and epidemiological perspectives in mercury emissions in small-scale gold mining in South Africa over the last 10 years. We searched the Web of Science Core Collection, PubMed and Scopus database, and ultimately included 40 articles. These included peer reviewed journal articles and conference proceedings, authentic book chapters, published and unpublished theses and reports and selected web references.

Mercury Use and Emissions

Mercury plays a central role in small-scale gold panning due to its unique ability to amalgamate the gold, allowing for the extraction of minute gold particles from the ore. However, despite efforts to recover mercury for reuse, a significant portion is lost during this process due to inefficiencies and improper handling practices.

The release of mercury into the environment occurs at multiple stages during small-scale gold panning operations. Firstly, mercury vaporises at relatively low temperatures, leading to atmospheric emissions during the amalgamation and gold burning stages (Global Mercury Partnership (GMP) 2018). The vapourised mercury can travel over long distances, contributing to regional pollution and its eventual inhalation by humans.

Environmental Health Impacts

The environmental impacts of mercury released during small-scale gold panning in South Africa can be far-reaching and multifaceted, with consequences for both terrestrial and aquatic ecosystems (Krabbenhoft 2004; Aldous et al. 2024). Once deposited onto terrestrial ecosystems, mercury can accumulate in soils and sediments, where it persists for extended periods, posing risks to plants, wildlife and soil microbial communities.

Exposure to mercury in ASGM communities in South Africa poses significant health risks, encompassing a spectrum of adverse effects on miners and adjacent communities. The populations most vulnerable include women of childbearing age, pregnant women and children (Bose-O'Reilly 2018). The health effects range from neurological disorders to reproductive and developmental impairments including impaired learning (see, e.g., Channa et al. 2013; WHO 2018; Charkiewicz et al. 2025; Fig. 4.1.); and underscore the urgent need for comprehensive interventions to mitigate exposure and protect the health of these vulnerable populations.

Socio-economic factors, limited access to healthcare and inadequate environmental regulations exacerbate their vulnerability to mercury toxicity, highlighting

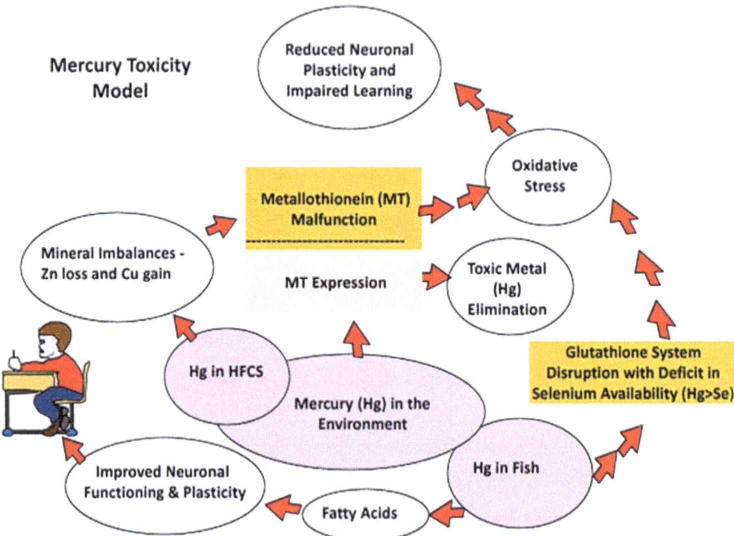

Fig. 4.1 The original mercury toxicity model. From Default et al. (2009). Reproduced with permission

the need for targeted interventions and public health initiatives to protect these at-risk groups.

Neurological Disorders

Chronic exposure to mercury vapour, a neurotoxic compound, can lead to a range of neurological disorders, including tremors, memory loss, cognitive impairments and motor dysfunction (Steckling et al. 2017; George et al. 2023). Mercury targets the central nervous system, disrupting neuronal function and neurotransmitter signalling, thereby manifesting in a variety of neurological symptoms among exposed individuals.

Respiratory Effects

Inhalation of mercury vapour during gold panning activities can result in respiratory symptoms such as coughing, chest tightness and respiratory distress (AGC) 2020). Prolonged exposure to mercury vapour may contribute to the development of respiratory conditions, including chronic bronchitis and interstitial lung disease, particularly in individuals with pre-existing respiratory conditions or compromised lung function.

Cardiovascular Effects

Emerging evidence suggests that chronic exposure to mercury may also impact cardiovascular health, increasing the risk of hypertension, myocardial infarction and cardiovascular mortality (Genchi et al. 2017; Hu et al. 2021). Mercury-induced oxidative stress and inflammation contribute to endothelial dysfunction and arterial stiffness, predisposing individuals to cardiovascular disease.

Reproductive and Developmental Impacts

Mercury exposure poses significant risks to reproductive and developmental health, with implications for both maternal and child health outcomes. Women of child-bearing age and pregnant women are particularly vulnerable to mercury toxicity, as mercury readily crosses the placental barrier, exposing the developing foetus to potential harm (Bose-O'Reilly 2018; Zinia et al. 2023). Adverse pregnancy outcomes, including preterm birth, low birth weight and developmental delays, have all been associated with maternal mercury exposure. However, current evidence suggests that dietary mercury exposure during pregnancy is unlikely to be a risk factor for low neurodevelopmental functioning in early childhood (see, e.g., Dack et al. 2022).

Children's Health

Children living in ASGM communities are at heightened risk of adverse health effects due to their developing physiology and increased susceptibility to environmental toxins (Rauh and Margolis (2016). Early-life exposure to mercury can disrupt neurodevelopmental processes, leading to cognitive deficits, learning disabilities and behavioural disorders (Allan-Blitz et al. 2022; Elumalai et al. 2023; Charkiewicz et al. 2025). Furthermore, children may experience elevated exposure to mercury through the consumption of mercury-contaminated breast milk, water and food sources, exacerbating their health risks (Bose-O'Reilly et al. 2020).

Addressing the Issues

Given the present seriousness of the mercury pollution situation, extensive research has been conducted throughout the world, not least in South Africa, to reduce the mercury content in the various environmental media as well as its removal in humans; and many countries have promulgated strict rules and regulations to control the growing concentrations of mercury in the environment (see, e.g., European Union 2024).

The imperative to address mercury emissions in small-scale gold panning has spurred the development of modern mining and ore processing methods, and implementation of various mitigation strategies aimed at reducing environmental contamination and protecting public health (see, e.g., Kung et al. 2024). Further efforts to reduce exposure will require drastic improvement in waste management practices, and promotion of sustainable mining practices. These imperatives would help minimise environmental degradation and safeguard ecosystem health (see, e.g., Bester 2023).

Mercury-Free Technologies

Mercury-free gold extraction technologies such as gravity concentration and cyanidation (see, e.g., Verbrugge et al. 2021; Keane et al. 2023), offer viable alternatives to traditional amalgamation methods, eliminating the need for mercury use in the gold extraction process. Gravity concentration techniques, including centrifugal concentrators and shaking tables, exploit differences in density to separate gold from other minerals, while cyanidation involves the use of cyanide to dissolve gold from the ore (Malone et al. 2023). By promoting the adoption of mercury-free technologies, mining communities can significantly reduce mercury emissions and minimise environmental contamination associated with small-scale Au panning.

Safer Mining Practices

In addition to technological interventions, the promotion of safer mining practices is essential for mitigating mercury emissions and promoting sustainable gold mining operations. This includes the implementation of best practices in ore processing, waste management and site remediation to minimise environmental impacts and protect human health. Safer mining practices may involve the use of protective equipment, such as respiratory masks and gloves, to reduce direct exposure to mercury vapour and dust during gold panning activities. Furthermore, proper ventilation and containment measures can help minimise atmospheric emissions and prevent mercury contamination of soil and water resources.

Incentive Programmes

Incentive programmes can play a crucial role in promoting the adoption of mercury-free technologies and safer mining practices among ASGM practitioners. Government subsidies, tax incentives and access to credit facilities can help offset the initial costs associated with investing in alternative technologies and equipment and accelerate the move towards formalisation. Additionally, certification schemes, such as Fairmined and Fairtrade Gold (Chuma 2021), provide market incentives for responsibly produced gold, encouraging miners to adhere to stringent environmental and social standards.

Policy and Regulatory Frameworks

Strong policy and regulatory frameworks are essential for driving the adoption of mercury-free technologies and promoting responsible mining practices within the small-scale Au-mining sector.

The South African Mercury Assessment (SAMA) Programme (Williams 2007; Matthews 2008) is incorporated in a White Paper drawn up in 2006, articulating the need to develop a framework for mercury research in South Africa. The research areas addressed in the SAMA Programme include (a) regulatory framework; (b) analytical methods; (c) source, speciation, fate, and transport; and d) impacts (ecological and human health). Aspects of the South African mercury assessment (SAMA) programme were reviewed in 2007 (e.g., Leaner 2007).

In South Africa, the Department of Mineral Resources and Energy (DMRE) is responsible for overseeing the mining sector and enforcing regulations related to ASGM activities (Chuma 2021); and has implemented measures to control mercury use and emissions, including licensing requirements, environmental impact assessments and monitoring and enforcement mechanisms (see: DMRE 2022). An analysis of the Minamata Convention and its implications for the regulation of mercury in South Africa is presented by Ross (2017). Additionally, the Government works closely with mining communities, industry stakeholders and civil society organisations to promote compliance with regulatory standards and support the transition to safer and more sustainable mining practices (see, e.g., Muthelo et al. 2022).

On 26 January 2024, the Government of South Africa approved the initiation of a Project by the Global Environment Facility (GEF) of the United Nations". To assist the Government of South Africa in the development of its National Action Plan for the Artisanal and Small-scale Gold Mining (ASGM) sector, raise awareness on the Minamata Convention, and to build al national capacity for the early implementation of the National Action Plan and the Minamata Convention" (Global Environment Facility (GEF) 2024).

International policy frameworks play a crucial role in addressing mercury emissions from ASGM activities, providing a regulatory framework for promoting environmentally sustainable practices and safeguarding public health (Aldous et al. 2024). International agreements, such as the "Minamata Convention on Mercury" (The United Nations Development Programme (UNDP) 2024), provide a framework for global cooperation and action to reduce mercury emissions from ASGM activities.

Capacity Building and Technical Assistance

In addition to regulatory measures, governments and international organisations provide capacity building and technical assistance to support ASGM communities in adopting mercury-free technologies and implementing best practices. Capacity building initiatives may include training programmes, workshops and knowledge sharing platforms aimed at empowering miners with the skills and knowledge needed to transition to safer and more sustainable mining methods. Technical assistance may involve the provision of funding, equipment and technical expertise to support the adoption of mercury-free technologies and the implementation of environmental management plans.

Stakeholder Engagement and Collaboration

Effective policy frameworks for addressing mercury emissions from ASGM activities require collaboration and engagement with a wide range of stakeholders, including government agencies, mining communities, industry associations and civil society organisations. Stakeholder engagement processes facilitate the development of inclusive and participatory policy measures that consider the perspectives and needs of all stakeholders. Collaboration between governments, industry stakeholders and civil society organisations is essential for implementing and enforcing regulatory measures, promoting transparency and accountability and fostering a culture of responsible mining practices.

Adsorption Research on Combatting Mercury Toxicity in Humans

During the last century, various techniques have been employed to remove mercury ions from different contaminated environmental media; but recent research has shown adsorption to be one of the most competitive and efficient methods, particularly when utilising suitable adsorbents. Among a wide array of materials used in adsorption trials, zeolite stands out as a promising, economical and environmentally friendly material. For the clean-up of mercury, the use of natural and engineered zeolites is also one of the most well-established adsorption technologies. The attractiveness of zeolite as an absorbent for toxic metals (re.: mercury) is bolstered by its well-defined micropore dimensions and a composition within a strict crystal lattice.

Among the different known structural varieties to date, *clinoptilolite* is the most abundant and commonly used natural zeolite. With its high cation-exchange capacity, large surface area and strong adsorption properties, studies in the field of toxicology and veterinary medicine have demonstrated that clinoptilolite is non-toxic and safe for use in both in vivo and in vitro applications (Senila and Cadar 2024). Additionally, their structural and compositional richness allows for fine-tuning of their adsorption properties to address a targeted separation. *However, despite its extensive use, many underlying action mechanisms of clinoptilolite in its natural or modified forms are still unclear, especially in humans* material (see, e.g., Selvam et al. 2018; Bulog et al. 2024).

Research on Zeolite Synthesis and Applications in South Africa

In South Africa, the field of zeolite synthesis and applications is still at its infancy, yet it is here that prospects for effective application, especially in the medical field, is thought to be at its highest level. The few available publications on South African zeolites deal with synthesis and characterisation (e.g., Diale et al. 2011; Musyoka et al. 2014; Mokgehie et al. 2020; Sinngu et al. 2022; Collins et al. 2023). In terms of applications, South African zeolite studies refer mainly to industrial applications and pollution studies, such as improvement of adsorption of zeolite effectiveness for environmental pollutants. For example, Wanyonyi et al. (2024) researched the capacity of South African heulandite zeolite to remove Pb^{2+} and Cd^{2+} ions from aqueous solution using batch experiments and molecular simulations studies. In *South Africa, more studies are needed that focus on geomedical (*in vivo, *human) applications as the primary target.*

Other Research Gaps and Suggestions for Future Work

Despite significant research efforts over the last decade, several knowledge gaps persist in our understanding of the dynamics of mercury emissions in ASGM and from other environmental sources in South Africa. These include the effectiveness of mitigation strategies in real-world settings, the socio-economic drivers of mercury use, the long-term health effects of chronic mercury exposure among ASGM communities and effective ways of mercury removal in humans.

Policy Recommendations

- There is a need for strengthening and enforcing existing regulations governing small-scale gold mining to minimise mercury emissions and protect environmental and public health (see: Madonsela et al. 2025; Manduna 2025).
- Provide incentives and support for the adoption of mercury-free technologies and sustainable mining practices among small-scale miners to help in combatting environmental health concerns.
- Fostering better collaborative ties between government agencies, mining communities, industry stakeholders and civil society organisations will enhance the development and implementation of effective policy measures.
- The promotion of community-based initiatives aimed at raising awareness about the health risks associated with mercury exposure and empowering miners with the knowledge and skills needed to adopt safer mining practices.
- Establishment of community-led monitoring and surveillance programs to track mercury emissions, assess environmental contamination and monitor health outcomes in ASGM communities.
- Expanding capacity building and training programmes that will provide miners with technical assistance, education and support in transitioning to mercury-free technologies and environmentally sustainable mining practices.
- Development of curriculum and training materials tailored to the needs and preferences of small-scale miners, emphasising practical skills, safety measures and environmental stewardship constitutes an important measure for mitigating health risks and related consequences of ASGM.
- Encouraging further research and innovation in the development of mercury-free technologies, alternative gold extraction methods and remediation techniques for mitigating mercury contamination in ASGM environments. Such work should examine the socio-economic impacts of transitioning to mercury-free technologies on small-scale mining communities, including implications for livelihoods, employment opportunities and socio-cultural dynamics (see, e.g., Chuma 2022).
- Conduct epidemiological surveys including longitudinal studies to assess the long-term health effects of chronic mercury exposure and evaluate the effectiveness of intervention strategies in reducing mercury-related morbidity and mortality.
- Foster international collaboration and knowledge sharing among other African countries with large ASGM sectors to exchange best practices, lessons learned and innovative solutions for addressing mercury emissions and promoting sustainable mining practices (see, e.g., Bester 2023).
- Advocate for increased funding and support from international organisations, donor agencies and development partners to scale up efforts in addressing mercury pollution in the South African ASGM communities.
- Finally, in March 2025, the Department of Forestry, Fisheries and the Environment (DFFE) (2024) of South African promulgated the "Regulations for the Management of Mercury" (Government Notice 6073) which are now in force. These regulations put in place "… comprehensive controls on mercury, including phasing out

mercury-added products, regulating manufacturing processes, and implementing mandatory registration and reporting requirements to ensure the environmentally sound management of mercury in the country".

Conclusion

Despite the availability of mitigation strategies, the adoption of mercury-free technologies and generally safer mining practices, ASGM communities in South Africa still face challenges related to affordability, technical feasibility, socio-economic factors and, most importantly, health concerns. Addressing these barriers requires a multi-faceted approach that integrates further technological innovation, capacity building, policy reform and stakeholder engagement to promote sustainable and responsible gold mining practices and mitigate the environmental and health impacts of mercury emissions.

This systematic review aimed at providing new contextual information (acquired from bibliographic references in the last 10 years for reducing or eliminating mercury emissions using emerging technologies. The study provides fresh insights into how the dynamics of mercury exposure can be better elucidated. Such an attainment would lead to improved diagnoses and therapy of conditions resulting from mercury exposures of miners and associated communities in alluvial gold mining centres, with particular reference to South Africa which hosts one of the largest ASGM sector in Africa. The study emphasises the profound environmental and public health hazards posed by mercury pollution, affecting not only the immediate ASGM communities but also the surrounding ecosystems.

Our study constitutes a background for future researchers seeking to gain a comprehensive understanding of the recent advancements in the field; and synthesises extant data on the complexities of mercury emissions and exposure in ASGM in South Africa. By equipping individuals with the knowledge and resources to navigate the dynamics of mercury pollution, mining communities in South Africa can actively participate in the safeguarding of their health and environment.

References

Aldous AR, Tear T, Fernandez LE (2024) The global challenge of reducing mercury contamination from artisanal and small-scale gold mining (ASGM): evaluating solutions using generic theories of change. Ecotoxicology 33:506–517. https://doi.org/10.1007/s10646-024-02741-3. Accessed 10 Apr 2024

Allan-Blitz LT, Goldfine C, Erickson TB (2022) Environmental and health risks posed to children by artisanal gold mining: a systematic review. SAGE Open Med 10:205031212210769936. https://doi.org/10.1177/20503121221076934. Accessed 10 Apr 2024

Artisanal Gold Council (AGC) (2020) Chemical hazards in the artisanal gold sector Impacts of mercury, cyanide and silica dust on human health and environment. https://www.planetgold.org/sites/default/files/Chemical-hazards-in-ASGM.pdf. Accessed 11 Apr 2024

Bester V (2023) Towards a sustainable artisanal gold mining sector in South Africa: proposed developmental initiatives. J Rural Stud 97:375–384. https://doi.org/10.1016/j.jrurstud.2022.12.029. Accessed 10 Apr 2024

Bose-O'Reilly S (2018) 1637b Health effects of mercury poisoning among miners and families in ASGM. Occup Environ Med 75:A242.2–A242. https://doi.org/10.1136/oemed-2018-ICOHabstracts.690

Bose-O'Reilly S, Lettmeier B, Shoko D, Roider G, Drasch G, Siebert U (2020) Infants and mothers' levels of mercury in breast milk, urine and hair: data from an artisanal and small-scale gold mining area in Kadoma/Zimbabwe. Environ Res 184:109266. https://doi.org/10.1016/j.envres.2020.109266. Accessed 23 Aug 2025

Bulog A, Pavelic K, Šutić I, Pavelic SK (2024) PMA-Zeolite: Chemistry and diverse medical applications. J Funct Biomater 15:296. https://doi.org/10.3390/jfb15100296. Accessed 07 Oct 2024

Channa K, Odland JØ, Kootbodien T, Theodorou P, Naik I, Sandanger TM, Röllin HB (2013) Differences in prenatal exposure to mercury in South African communities residing along the Indian Ocean. Sci Total Environ 463–464:11–19. https://doi.org/10.1016/j.scitotenv.2013.05.055. Accessed 10 Apr 2024

Charkiewicz AE, Omeljaniuk WJ, Garley M, Nikliński J (2025) Mercury exposure and health effects: what do we really know? Int J Mol Sci 26:2326. https://doi.org/10.3390/ijms26052326. Accessed 10 Apr 2024

Chuma M (2021) Reframing artisanal and small-scale gold mining as a livelihood strategy and the role of law in constituting livelihood assets. Thesis presented for the Degree of Doctor of Philosophy (School of Law) in the Faculty of Commerce, Law and Management at the University of the Witwatersrand, Johannesburg. https://wiredspace.wits.ac.za/server/api/core/bitstreams/725c1462-b42d-441d-8f7d-c250f7251b48/content. Accessed 10 Apr 2024

Chuma M (2022) A Closer look on the relevant laws ring-fencing key livelihood assets employed in artisanal and small-scale gold mining (ASGM) in South Africa. https://ssrn.com/abstract=4243415 or https://doi.org/10.2139/ssrn.4243415. Accessed 10 Apr 2024

Collins AC, Strydom CA, Matjie RH, Bunt JR, van Dyk JC (2023) Production of sodium-based zeolites and a potassium-containing leach liquor by alkaline leaching of South African coal fines ash. J Southern Afric Inst Min Metall 123(3):145–155. https://doi.org/10.17159/2411-9717/1167/2023. Accessed 26 oct 2024

Dack K, Fell M, Taylor CM, Havdahl A, Lewis SJ (2022) Prenatal mercury exposure and neurodevelopment up to the age of 5 Years: a systematic review. Int J Environ Res Public Health 19(4):1976. https://doi.org/10.3390/ijerph19041976. Accessed 11 Apr 2024

Department of Forestry, Fisheries and the Environment, Government of the Republic of South Africa (DFFE) (2025) Regulations for the management of mercury in South Africa. Government Notice No. 6073. National Environmental Management Act No. 107 of 1998. https://media.lawlibrary.org.za/media/legislation/302151/source_file/6ba18a138ca144a2/regulations-for-the-management-of-mercury-in-south-africa-2025.pdf. Accessed 22 Aug 2025

Department of Mineral Resources and Energy of South Africa (DMRE) (2022) Artisanal and small-scale mining policy, final vi; no. 46124. https://www.gov.za/sites/default/files/gcis_document/202203/46124gon1938.pdf. Accessed 03 Jun 2025

Diale PP, Muzenda E, Zimba J (2011) A study of South African natural zeolites properties and applications. In: Proceedings of the world congress on engineering and computer science, vol. 2. WCECS October 19–21, San Francisco, USA, pp 19–21. https://hdl.handle.net/10210/10669. Accessed 26 Oct 2024

Dufault R, Schnoll R, Lukiw WJ, Leblanc B, Cornett C, Patrick L, Wallinga D, Gilbert SG, Crider R (2009) Mercury exposure, nutritional deficiencies and metabolic disruptions may affect learning in children. Behav Brain Funct: BBF 5(44). https://doi.org/10.1186/1744-9081-5-44. Accessed 08 Aug 2025

Elumalai V, Sujitha SB, Jonathan MP (2023) Mercury pollution on tourist beaches in Durban, South Africa: a chemometric analysis of exposure and human health. Mar Pollut Bull 180:113742. https://doi.org/10.1016/j.marpolbul.2022.113742

European Union (2024) Towards a mercury-free environment: 'Revised Mercury Regulation' enters into force. https://environment.ec.europa.eu/news/revised-mercury-regulation-enters-force-2024-07-30_en. Accessed 04 Oct 2024

Genchi G, Sinicropi MS, Carocci A, Lauria G, Catalano A (2017) Mercury exposure and heart diseases. Int J Environ Res Public Health 14(1):74. https://doi.org/10.3390/ijerph14010074. Accessed 21 Sept 2024

George J, Sadiq E, Moola I, Maharaj S, Mochan A (2023) Informal gold miners with mercury toxicity: Novel asymmetrical neurological presentations. South African Med J = Suid-Afrikaanse tydskrif vir geneeskunde 113(12):20. https://doi.org/10.7196/SAMJ.2023.v113i12.1127. Accessed 10 Apr 2024

Global Environment Facility (GEF) (2024) National action plan for the artisanal and small-scale gold mining (ASGM) sector in South Africa. GEF Project ID: 11494. https://www.thegef.org/projects-operations/projects/11494. Accessed 21 Sept 2024

Global Mercury Partnership (GMP) (2018) Technical background report for the global mercury assessment 2018. Chapter 3 E-annex: methodology for estimating mercury emissions to air and results of the 2015 global emissions inventory. https://wedocs.unep.org/bitstream/handle/20.500.11822/29832/gma_annexch3.pdf?sequence=1&isAllowed=y. Accessed 10 Apr 2024

Hu XF, Lowe M, Chan HM (2021) Mercury exposure, cardiovascular disease, and mortality: a systematic review and dose-response meta-analysis. Environ Res 193:110538. https://doi.org/10.1016/j.envres.2020.110538. Accessed 10 Apr 2024

Keane S, Bernaudat L, Davis KJ, Stylo M, Mutemeri N, Singo P, Twala P, Mutemeri I, Nakafeero A, Etui ID (2023) Mercury and artisanal and small-scale gold mining: review of global use estimates and considerations for promoting mercury-free alternatives. Ambio 52(5):833–852. https://doi.org/10.1007/s13280-023-01843-2. Accessed 11 Apr 2024

Krabbenhoft DP (2004) Methylmercury contamination of aquatic ecosystems: a widespread problem with many challenges for the chemical sciences. In: Norling P, Wood-Black F, Masciangioli TM (eds) Water and sustainable development: opportunities for the chemical sciences: National Research Council (US) chemical sciences roundtable. A workshop report to the chemical sciences roundtable. National Academies Press, Washington DC, USA. https://www.ncbi.nlm.nih.gov/books/NBK83731/. Accessed 10 Apr 2024

Kung HC, Wu CH, Huang BW, Chang-Chien GP, Mutuku JK, Lin WC (2024) Mercury abatement in the environment: insights from industrial emissions and fates in the environment. Heliyon 10(7):e28253. https://doi.org/10.1016/j.heliyon.2024.e28253. Accessed 21 Sept 2024

Landrigan P, Bose-O'Reilly S, Elbel J, Nordberg G, Lucchini R, Bartrem C, Grandjean P, Mergler D, et al (2022) Reducing disease and death from Artisanal and Small-Scale Mining (ASM)—the urgent need for responsible mining in the context of growing global demand for minerals and metals for climate change mitigation. Environ Health 21:78. https://doi.org/10.1186/s12940-022-00877-5. Accessed 21 Sept 2024

Leaner J (2007) Focus on CSIR research in pollution waste: South African mercury assessment (SAMA) programme. 2007 Stockholm World Water Week, 13–17 August 2007. p 1. https://researchspace.csir.co.za/items/f39954d0-c971-42a3-a370-80ea846edf21. Accessed 21 Sept 2024

Ledwaba PF (2017) The status of artisanal and small-scale mining sector in South Africa: tracking progress. J Southern Afric Inst Min Metall 117(1):33–40. https://doi.org/10.17159/2411-9717/2017/v117n1a6. Accessed 21 Sept 2024

Madonsela BS, Maphanga T, Grangxabe XS (2025) Environmental degradation from Zama-Zama illegal mining in South Africa: policy implementation and governance challenges. Sustainability 17:3418. https://doi.org/10.3390/su17083418. Accessed 22 Aug 2025

Makhetha E (2023) 'Zama Zama' and leftovers: the recycling of ore in abandoned gold mines in South Africa. Extract Indus Soc 14(2):101272. https://doi.org/10.1016/j.exis.2023.101272. Accessed 21 Sept 2024

Malone A, Figueroa L, Wang W, Smith NM, Ranville JF, Vuono DC, Alejo Zapata FD, Morales Paredes L, Sharp JO, Bellona C (2023) Transitional dynamics from mercury to cyanide-based processing in artisanal and small-scale gold mining: social, economic, geochemical, and environmental considerations. Sci Total Environ 898:165492. https://doi.org/10.1016/j.scitotenv.2023.165492. Accessed 11 Apr 2024

Manduna K (2025) Governance at the margins of the state: contextualising the spatial and temporal realities of illegal mining in contemporary South Africa. Extract Indus Soc 23:101694. https://doi.org/10.1016/j.exis.2025.101694. Accessed 22 Aug 2025

Matthews S (2008) Mercury levels in SA water resources probed. Water Wheel 7(1). Sabinet. https://hdl.handle.net/10520/EJC115674. Accessed 21 Sept 2024

Mokgehie TMA, Gitari WM, Tavengwa NT (2020) Synthesis and characterization of zeolites produced by ultrasonication of coal fly ash/NaOH slurry filtrates. South Afric J Chem 73:64–69. https://doi.org/10.17159/0379-4350/2020/v73a10. Accessed 26 Oct 2024

Musyoka NM, Missengue R, Kusisakana M, Petrik LF (2014) Conversion of South African clays into high quality zeolites. Appl Clay Sci 97–98:182–186. https://doi.org/10.1016/j.clay.2014.05.026. Accessed 03 Nov 2024

Muthelo L, Mothiba TM, Malema NR, Mbombi MO, Mphekgwana PM (2022) Exploring occupational health and safety standards compliance in the South African Mining Industry, Limpopo Province, using principal component analysis. Int J Environ Res Public Health 19(16):10241. https://doi.org/10.3390/ijerph191610241. Accessed 10 Apr 2024

Nhlengetwa K, Hein KA (2015) Zama-Zama mining in the Durban Deep/Roodepoort area of Johannesburg, South Africa: An invasive or alternative livelihood? Extract Indus Soc 2(1):1–3. http://www.sciencedirect.com/science/article/pii/S2214790X14000501. Accessed 10 Apr 2024

Rauh VA, Margolis AE (2016) Research review: environmental exposures, neurodevelopment, and child mental health—new paradigms for the study of brain and behavioral effects. J Child Psychol Psychiatry 57(7):775–793. https://doi.org/10.1111/jcpp.12537. Accessed 22 Aug 2025

Ross JC (2017) Thesis: A analysis of the minamata convention on mercury and its implications for the regulation of mercury in South Africa. Master of Laws in Environmental Law, University of KwaZulu-Natal, Pietermaritzburg. http://hdl.handle.net/10413/15879. Accessed 10 Apr 2024

Selvam T, Schwieger W, Dathe W (2018) Histamine-binding capacities of different natural zeolites: a comparative study. Environ Geochem Health 40(6):2657–2665. https://doi.org/10.1007/s10653-018-0129-5. Accessed 26 Oct 2024

Senila M, Cadar O (2024) Modification of natural zeolites and their applications for heavy metal removal from polluted environments: challenges, recent advances, and perspectives. Heliyon 10(3):e25303. https://doi.org/10.1016/j.heliyon.2024.e25303. Accessed 22 Oct 2024

Sinngu F, Ekolu SO, Naghizadeh A, Quainoo HA (2022) Experimental study and classification of natural zeolite pozzolan for cement in South Africa. J South Afric Instit Civil Eng 64(4):2–15. https://doi.org/10.17159/2309-8775/2022/v64n4a1. Accessed 03 Nov 2024

Steckling N, Tobollik M, Plass D, Hornberg C, Ericson B, Fuller R, Bose-O'Reilly S (2017) Global burden of disease of mercury used in artisanal small-scale gold mining. Ann Glob Health 83(2):234–247. https://doi.org/10.1016/j.aogh.2016.12.005. Accessed 11 Apr 2024

Street RA (2015) Commentary: emission accomplished: formal and informal mercury sources in South Africa. Clean Air J 25(2):9–11. https://doi.org/10.10520/EJC183712. Accessed 10 Apr 2024

The United Nations Development Programme (UNDP) (2024) Minamata convention. https://www.undp.org/chemicals-waste/chemicals-and-waste/minamata-convention. Accessed 10 Apr 2024

The United Nations Environment Programme (UNEP) (2013) The minamata convention on mercury. UNEP–UN Environment Programme. Accessed 21 Sept 2024

Verbrugge B, Lanzano C, Libassi M (2021) The cyanide revolution: efficiency gains and exclusion in artisanal- and small-scale gold mining. Geoforum 126:267–276. https://doi.org/10.1016/j.geoforum.2021.07.030

Wanyonyi FS, Orata F, Mutua GK, Odey MO, Zamisa S, Ogbodo SE, Maingi F, Pembere A (2024) Application of South African heulandite (HEU) zeolite for the adsorption and removal of Pb^{2+}

and Cd^{2+} ions from aqueous water solution: experimental and computational study. Heliyon 10(14):e34657. https://doi.org/10.1016/j.heliyon.2024.e34657. Accessed 26 Oct 2024

Williams C (2007) The South African mercury assessment (SAMA) programme. http://www.waternet.co.za/samercury/res_frame.html. Accessed 21 Sept 2024

World Health Organization (WHO) (2018) Assessment of prenatal exposure to mercury: human biomonitoring survey. The first survey protocol. Regional Office for Europe; 2018. Licence: CC BY-NC-SA 3.0 IGO. World Health Organization, Copenhagen. https://iris.who.int/bitstream/handle/10665/334181/WHO-EURO-2020-1069-40815-55163-eng.pdf. Accessed 13 Aug 2025

Zinia SS, Yang K-H, Lee EJ, Lim M-N, Kim J, Kim WJ, Ko-CHENS Study Group (2023) Effects of heavy metal exposure during pregnancy on birth outcomes. Sci Rep 13:18990. https://doi.org/10.1038/s41598-023-46271-0. Accessed 21 Sept 2024

Chapter 5
Geogenic Dust in the African Environment: Fate of Inhaled Particles from Mining, Mineral Processing and Related Processes

T. C. Davies, B. Mohan George, and M. Kgware

ABSTRACT Air particulates, commonly known as PM, constitute an everyday air pollution issue arising primarily from various sources, many of them natural or *geogenic*. Harmful respiratory conditions affecting the health of adolescents residing near areas known for airborne contaminants in Sub-Saharan Africa remain underresearched; however, certain studies indicate that *Lower Respiratory Infections* rank second in the most recent listing of *Leading Causes of Death in Africa*. Living near locations where silica-rich ores are mined or processed and where coal combustion takes place, can contribute significantly to respiratory issues among people residing in nations endowed with these resources through exposure to emitted dust. In African coal mines, particulate matter composed of respirable quartz and coal dust is generated as part of normal operations, and poses significant risks to miners' health and safety. Current research highlights various ailments encompassing respiratory conditions like silicosis, pulmonary fibrosis, asbestos-related illnesses, *coal worker's pneumoconiosis* (CWP), heart disease and fatalities under severe circumstances. These maladies have been recognised as occupational illnesses for many years because they result from inhaling silica dust during mining operations involving rock formations containing silicate minerals. Materials such as asbestos fibres extracted from mines are processed into industrial products. Despite several years of study of the clinical outcomes of dust exposure from these operations, our understanding of the pathogenesis of dust-related diseases in Africa remains incomplete; and research on how immune system responses contribute to the development of respiratory illnesses continues to this day. Through an abbreviated meta-analysis approach, our study investigates contemporary African literature focussing primarily on work done

T. C. Davies (✉) · B. M. George
Faculty of Applied and Health Sciences, Mangosuthu University of Technology, Umlazi, KwaZulu Natal Province, Republic of South Africa
e-mail: davies.theophilus2025@yahoo.com

M. Kgware
Community Health Studies, Durban University of Technology, Durban, KwaZulu Natal Province, Republic of South Africa

T. C. Davies (ed.), *Recent Advances in Medical Geology Research in Africa: A Decadal View*, SpringerBriefs in Earth System Sciences,
https://doi.org/10.1007/978-3-032-13754-8_5

"

after 2010 on dust exposures and their associated toxic effects. This analysis aims at elucidating the underlying causes and pathways leading to diseases caused by dust accumulation; consequently, it guides the development of efficient strategies for controlling dust sources as well as interventions aimed at slowing down illness escalation.

Keywords Dust exposure · Silicate ores · Asbestos mining and processing · Pneumoconiosis · Addressing the issues

Introduction

African populations continue to be increasingly concerned about exposure to particulate matter (PM) in ambient air, especially in urban areas. While the effects of dust exposure from industrial processes on human health are well-documented and widely regulated (e.g., Tian et al. 2024; Huboyo et al. 2025), dust from geological sources and less visible fine particles have received much less research attention. This is probably partly because research is in many cases at an early stage, and due to the realistic problems of observing, monitoring and sampling dispersed aerosols within a complexly dynamic atmospheric column (Yu and Zahidi (2023). *Furthermore, large knowledge voids exist in our ability to predict future trends in Sahara dust emissions and model-projected atmospheric circulation styles* (see Section on: "The Sahara Dust Plume", this chapter).

It is hoped that the overview presented here will address many of the perceived gaps in understanding; and the updated list of references provided will strengthen the search process of researchers who want to further address concerns about geological dust emissions and risks in Africa. This discussion will further contribute to the feasibility of formulating concrete proposals for remedial interventions to (reduce) the harmful effects of geogenic dust.

Methodology

In this study, we carried out a mini-systematic review of recent studies that reported epidemiological results on the geogenic dust releases and exposures of the African population, especially those living in close proximity to silica, asbestos and coal mining hubs.

The review followed the PRISMA (Preferred Reporting Items for Systematic Reviews and Meta-Analyses) Protocol (Page et al. 2021) and referred to extant, historical long-term data held within clinical registries on respiratory diseases in Africa, using South Africa in our case descriptions, South Africa being one of the world's foremost producers of silicate and asbestos ores. Our discourse further examines the extent to which advances in reducing hazards in African mines have been achieved.

We searched the databases of PubMed, Scopus and the Web of Science Core Collection, producing an assortment consisting of a total of forty articles. These articles included peer-reviewed journal papers, real book chapters, conference proceedings, both published and unpublished theses and reports, and selected web-based references.

Sources of Geogenic Dust

A substantial portion of particulate load in African cities consists of particles from geogenic sources (Davies 2008; Doumbia et al. 2023). The toxicity to people of these forms of dust may be even worse than that from dust derived from other sources. Major contributors to African airborne dust include (i) Harmattan winds carrying substantial quantities of Sahara Desert dust, stretching through West Africa between November and February; occasionally mixed with radioactive material; (ii) dust from industrial activities such as mining and quarrying; (iii) processing facilities; (iv) landfill operations; (v) unsealed pathways; (vi) exposed surfaces; (vii) cultivated areas; (viii) reclaimed lands; (ix) building zones; and (x) wood chips emissions, encompassing various burning processes. *According to* Formenti et al. (2011), *our understanding of how dust characteristics evolve throughout transportation remains incomplete.* In Africa, current research focuses on understanding how minerals in dust particles vary as they are emitted into the atmosphere and then evolve throughout transportation processes. The significant research attention given to these areas is as a result of their importance for climate studies (see: Fernandes et al. (2023); Formenti and Biagio (2024).

Sources of Air Pollution

Air pollutants can be grouped into three types based on where they come from: primary, which are directly released into the air; secondary, which form when pollutants react in the atmosphere; and re-emission, which happens when primary or secondary pollutants are released again. This way of grouping helps us understand how and where pollution comes from; which in turn helps us see how much people are exposed to different types of pollution (IARC WGECRH 2016). Using solid fuels at home is also a major reason for high levels of $PM_{2.5}$ in the air in many African countries (Health Effects Institute (HEI) 2022).

　　　　　　　　　　　　　　　　　　　　　　　　　　　　　　　　　　　

Categories of Particulate Matter

There are three main types of particulate matter: coarse particles (PM_{10}), fine particles ($PM_{2.5}$) and ultrafine particles ($PM_{0.1}$). These different sizes come from different sources and can affect health in different ways. Fine particles ($PM_{2.5}$) are tiny and can be inhaled into the lungs, and they are among the most well-known air pollutants. These particles are usually 2.5 micrometres or smaller. Ultrafine particles ($PM_{0.1}$), which are even smaller, are also a big health concern because they are found in large amounts in the air (Schraufnagel 2020). Fine particles ($PM_{2.5}$) can come from both natural (geogenic) and man-made sources (Health Effects Institute (HEI) 2022).

In Africa, the sources of $PM_{2.5}$ pollution can be different from one country to another and even within the same region (Tessum et al. 2022). Natural sources of particulate matter include windblown dust, sea spray, wildfires and ash and dust clouds from volcanic eruptions. Man-made sources include burning fossil fuels and biofuels, transportation, industrial activities, farming and burning waste.

Dust Generated from Mining, Ore Processing and Related Activities

Mine dust comes from mining and processing of ores as well as other related activities. Dust forms when rocks break apart due to impact during operations like drilling, crushing, moving, grinding, blasting and stockpiling minerals. The makeup of this dust depends on where the minerals are found and closely matches the soil in that area. The dust is named based on the main mineral component. Examples include silica dust, asbestos dust and coal dust. The type of dust is a major health issue. Dust comprising harmful mineral particles (e.g., asbestos; lead, uranium) can harm mine workers causing diseases like silicosis and lung cancer from crystalline silica, and coal workers' pneumoconiosis (CWP) from coal dust. (See sections: "Health Effects of Silica Dust", Chapter 2 and "Health Effects of Silica Dust", this chapter). Mine dust also harms the environment. It damages plants and landscapes, affects weather and reduces air quality, which leads to major environmental problems. The harmful substances in mine dust can pollute nearby rivers, farmland and crops. This poses a serious risk to the water and food supplies of nearby communities, and these people may suffer from various diseases due to exposure to mine dust. (See Section on: "Health Effects of Exposure to Mineral Dust," this chapter).

The Sahara Dust Plume

A huge amount of dust, estimated to be around a million tons, is released from the Sahara Desert each year into the atmosphere. This dust forms what is known as the

Sahara Dust Plume (SDP) and moves across the North Atlantic Ocean. TheSDP plays a key role in increasing the levels of particulate matter in parts of North Africa, the Middle East and some areas in western Sub-Saharan Africa. It can also reach more distant regions such as the Caribbean, North America and Europe (The World Meteorological Organisation (WMO) 2023).

Transport of African Dust

Soil that is exposed to the wind can be eroded, creating airborne particles that range from fine dust to larger silt-like material. Once in the air, dust can travel long distances before settling on land or water. For example, the SDP can reach the Americas and Europe, depending on weather patterns (see, e.g., Gomes et al. 2022). Dust can also come from specific sources such as volcanic eruptions, fires and even atmospheric nuclear tests. Areas with unstable materials, like tips and lagoons that contain mineral waste, can also produce dust when dry. In rural parts of Africa, most roads are made of a material called *murram*, which is 'not tarred' or tarmac. This type of road can mix with dust that has harmful metals like lead and arsenic, as well as other harmful substances, including

pathogens found in pesticides (see: Degrendele et al. 2022). This dust gets blown into the air and is easily inhaled, which can lead to problems like asthma and other respiratory issues.

Asbestos

Asbestos is a group name for six natural fibrous minerals made up of silicate materials. These include amosite (also known as grunerite), crocidolite (riebeckite), tremolite, anthophyllite and actinolite, which all fall underthe amphibole group. The sixth type, chrysotile, has a different structure and is classified as 'serpentine mineral' (see Section on "Asbestos Dust" in Chapter 2 of this Volume).

These minerals were widely used in industries during the last century because of their unique and useful properties. They were used in various products like insulation, pipe covering, roof materials and brake pads. Today, asbestos is mostly known for its harmful effects. It can cause serious diseases like asbestosis, which is a lung condition that can be very damaging and even deadly. It is also linked to cancers such as lung cancer and mesothelioma, which can develop many years after exposure (Refer to the section on "Health Effects of Asbestos Exposure" in this chapter). It is important to note that both smoking and asbestos exposure together greatly increase the risk of lung cancer, but not mesothelioma (Klebe et al. 2019).

For over a century, South Africa was one of the world's top three asbestos producers and the main producer of amphibole asbestos (Braun and Kisting 2006). Large amounts of crocidolite and amosite are found in metamorphosed rock layers

Fig. 5.1 Asbestos cement waterpipes: A health hazard. *Credit* Helen Suzman Foundation. https://hsf.org.za/publications/hsf-briefs/asbestos-cement-waterpipes-a-health-hazard (Accessed 16.10.2024). Permission for reuse granted on 10.07.2025 by Helen Suzman Foundation (HSF) as the original *source* Courtesy: Yvette Odinga, Office Administrator, HSF: Yvette@hsf.org.za; info@hsf.org.za

from the Precambrian Era, specifically in the Transvaal Supergroup in South Africa. About 75% of the asbestos produced was used in making cement pipes that contain asbestos (see Fig. 5.1). Due to unfair and harsh working conditions, former workers and people who lived near asbestos mines have faced, and still face, a heavy and often underreported burden of diseases related to asbestos.

Restrictions on Asbestos Use

Asbestos processing and use are slowly being limited or completely banned in many countries, including the whole European Union, which stopped allowing asbestos in 2005 (The European Union (EU) 2022). There is also an updated EU Directive from 2023 called 'Amended EU Directive 2023/2668' issued in 2023 (CEU EP 2023). However, asbestos dust continues to be a big danger to workers and people in the EU. In 2019 alone, over 70,000 workers in the EU died because of past exposure to asbestos (The European Union (EU) 2022). In Africa, the situation is even worse. In South Africa, for example, "Asbestos Regulations" were introduced in 2002, which outlined how to work with and remove asbestos-containing materials (Braun and Kisting 2006). Then, in 2008, South Africa completely banned the use of asbestos (ECA 2008). Even though there is strong proof that asbestos is harmful to workers, it is still being processed and used in many other African countries and in several developing countries outside Africa (Frank 2020; Thives et al. 2022; Frank and Zandwijk 2024).

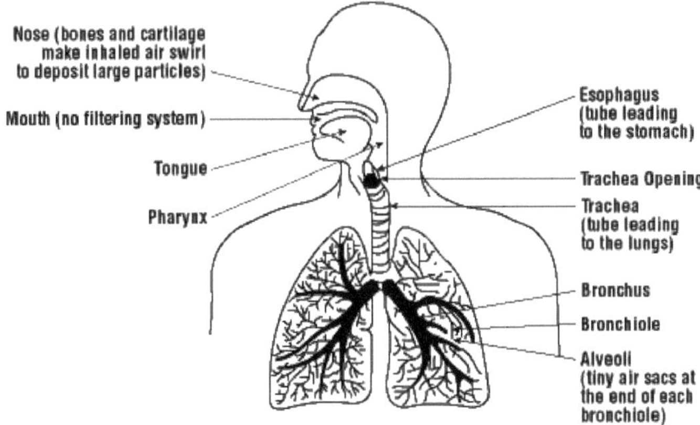

Fig. 5.2 Dust inhalation and its pathway through the airways. *Credit* Canadian Centre for Occupational Health and Safety (CCOHS). *From* CCOHS (2025). Chemicals and Materials: What are the effects of dust on the lungs? https://www.ccohs.ca/oshanswers/chemicals/lungs_dust.html (Accessed 18.10.2024). Reproduced with permission

Health Effects of Exposure to Mineral Dust

The health impacts of breathing in Earth materials depend on many factors related to the material itself. These include the type of material–like solid, liquid, gas or even pathogens—*viz.,* the chemical makeup of the material, how much of it is in the air you breathe, how well it dissolves and reacts with the fluids in your lungs and for solid particles, their shape and size (Plumlee et al., 2006).

The size and shape of solid particles affect how deep they can go into the respiratory system and how easily they can be removed by the body's natural cleaning processes (Fig. 5.2). Most dust particles trapped in the trachea and bronchioles are cleared from the body within a few hours, but a very small amount can get into the cells and tissues (McClellan, 2000). Long-term exposure to airborne dust over many years or decades can lead to various breathing problems, including higher rates of total, cardiovascular and infant deaths (e.g., Pouri et al. 2024).

Health Effects of Silica Dust

The term used to describe lung disease caused by breathing in inorganic dusts is *pneumoconiosis.* It refers to dust building up in the lungs and the body's reaction to its presence. This can also lead to other conditions like congestive heart failure and illnesses caused by germs that are linked to dust or soil, such as Valley Fever and naturally occurring anthrax (Tong et al. 2023). *Silicosis*, a form of pneumoconiosis linked to miners, is caused by certain types of crystalline silica, specifically quartz, tridymite and cristobalite. Silicosis and other diseases related to respirable crystalline silica, such as tuberculosis, have been significant health problems in South Africa for a long time (Brouwer and Rees 2020; Ehrlich et al. 2021; Maboso et al. 2023). *Despite extensive research since the 1950s on the harmful effects of silica, there is still no clear agreement among physicians about how crystalline silica affects the body at the molecular level* (see, e.g., Hoy and Chambers 2020; Skuland et al. 2020; Marques Da Silva et al. 2022).

Silicon is a component of *diatomite*, also known as *diatomaceous earth*. This material is mined in countries like Algeria, Ethiopia, Kenya and South Africa and is used in a variety of applications, such as filters, fillers, in anaerobic digestion (e.g., Wei et al. 2021), as an adsorbent (Zhang et al. 2023), in nanomedicine for treating colorectal cancer (Tramontano et al. 2023), and even as a mild abrasive (Zuluaga-Astudillo et al. 2023). Recently, Mwangi (2023) has highlighted several health benefits of diatomite, such as lowering blood cholesterol and being essential for tendons, cartilage, blood vessels and bones. However, exposure to diatomite dust during processing and handling can be very dangerous to people. It can lead to serious, even fatal, illnesses for those working with it or living near industrial sites, or even in the same household as someone who is exposed. During calcining, highly reactive forms of crystalline silica (such as tridymite and cristobalite) are produced, leading to silicosis (Nattrass et al. 2015; Nelson 2023; Yi et al. 2023). The illnesses include pulmonary fibrosis (silicosis), secondary heartdisease, lung cancer and bronchitis.

Coal Workers' Pneumoconiosis

Coal workers' pneumoconiosis (CWP), or *black lung disease*, is a long-term lung condition caused by breathing in large amounts of coal dust, leading to lung inflammation and scarring (fibrosis) (see under "Health Effects of Silica Dust," Chapter 2 of this Volume). In Africa, CWP is a major health problem, with a noticeable increase in its occurrence in North and Central Sub-Saharan Africa between 1990 and 2019 (Sun et al. 2024; Wang et al. 2024).

The disease is most commonly found among coal miners due to prolonged exposure, although other factors, such as the silica content in the dust and genetic predisposition, can also increase the risk. The disease develops over time, with symptoms such as coughing, shortness of breath, chest tightness and more severe complications like progressive massive fibrosis (PMF), often after years or even decades of exposure.

There is no cure for CWP, but there are various management strategies available, including oxygen therapy, pulmonary rehabilitation and quitting smoking (Coleman et al. 2023). South Africa, a major coal-producing country, has seen many documented cases of CWP, highlighting the need for ongoing health monitoring and prevention efforts for coal miners.

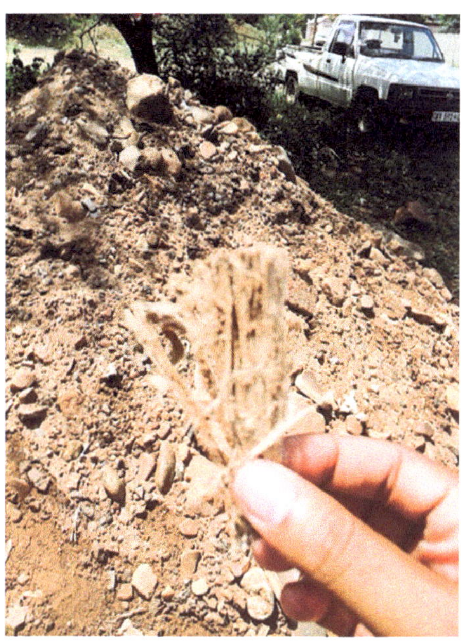

Fig. 5.3 Asbestos fibres around the streets of Penge village, Limpopo Province, north of Gauteng, South Africa. Authors' own image

Health Effects of Asbestos Exposure

Exposure to asbestos can lead to various lung and pleural diseases, including mesothelioma, lung cancer, asbestosis and pleural plaques (see, for example, Markowitz 2022; Caceres and Venkata 2023).

As of 2017, there were over a million homes in South Africa with asbestos cement roofs (Rees and Phillips 2017), which can release asbestos dust onto roadsides (Fig. 5.3). It is estimated that, due to the long latency period—often more than 20 years—the number of mesothelioma cases in South Africa is expected to rise because of the high production of asbestos in the 1990s (Muteba 2018).

Health Effects of the Sahara Dust Plume

From an environmental health standpoint, the Saharan Dust Plume (SDP) worsens air quality, leading to significant health risks for people, especially those with existing lung problems. Recent studies have reported health issues like respiratory and cardiovascular diseases (Rowan et al. 2024; Papatheodorou and Achilleos 2025) and, in severe cases, even death (see: Wang et al. 2020). These health problems are more common during the main dust emission season, which runs from November to March (Georgakopoulou et al. 2024). The dust storms from the Sahara also contain harmful inorganic particles and other pathogens, which can cause stress on the respiratory system when inhaled (Bredeck et al. 2023).

People who are more vulnerable, such as the elderly, young children and those with chronic lung conditions are at a greater risk and may suffer serious health complications due to dust exposure. Common symptoms that are often seen include difficulty breathing, wheezing, coughing and irritation of the nose and eyes.

Georgakopoulou et al. (2024) have compiled the existing scientific evidence about the effects of Saharan dust on respiratory health and have investigated the environmental movement of dust, the physical and chemical composition of dust particles and the biological impact of these particles on human health. Li et al. (2025) have also explored the gaps in health research related to Sahara dust and emphasised the need for more international, collaborative and multidisciplinary studies.

Health Effects of Other Mineral Dusts from Mining

Other types of mineral dusts that can cause harmful effects on the respiratory system, but are not yet confirmed as cancer-causing agents, include kaolinite (kaolinosis), talc (talcosis) and iron oxides (siderosis) (see: Verlynde et al. 2018; Nemery 2022; Hoskin et al. 2024).

It has been challenging to determine the exact causes of diseases from inhaling these minerals, mainly because of the complexity of epidemiological studies that involve exposure to other minerals like quartz and asbestos (Ross et al. 1993; Huang et al. 2006; Fubini and Feneglio 2007). Volcanoes produce particles that range from nanometres to micrometres in size, including crystalline silica, which may be harmful. These volcanic ash particles are broken apart during eruptions, creating fresh and reactive surfaces that are not typically found in the natural environment.

The Intersection of Climate Change and Dust Pollution

The two issues of air pollution and climate change are closely connected, as the sources of emissions that contribute to both are often the same, *viz.*, the burning of fossil fuels and biofuels (Fuller et al. 2022). Climate change affects air quality, which in turn can lead to negative health outcomes. Changes in weather patterns influence air quality by increasing and spreading air pollutants, such as fine particulates, ground-level ozone, wildfire smoke and dust. *There is still a lot more research needed to fully understand how dust impacts the climate. Long-term data is essential in remote areas, such as the dust source regions, for making informed long-term predictions about air pollution and climate.*

Recently, Ofremu et al. (2025) gave a detailed review of the link between climate change, air pollutants and human health; and provided useful insights into adaptation and mitigation strategies for decision-makers, public health experts and researchers working on the effects of climate change on human health.

Value-Added Contributions Made

A review of the literature from the past decade shows little additional research on the effects of dust exposure in Africa beyond the comprehensive global review of the toxicity of mineral dusts to humans conducted by Fubini and Feneglio in 2007. This Chapter, therefore, offers an updated and critical review of the current knowledge on geogenic dust pollution and its effects in Africa. The summary provides a good foundation, up-to-date references and a strong structure for exploring the various knowledge gaps (some of which are italicised in the text) related to dust exposure, the development of diseases, control and diagnosis, which still need further research.

Conclusion

Research into controlling dust and reducing its health risks is essential for the African society. In the Continent's mining and mineral processing industry, diseases caused by dust exposure and its consequences are among the leading causes of illness and death. Dust is a critical part of Earth's systems, influencing atmospheric, oceanic, biological, terrestrial and human processes, especially human health. *The goal of this work has been to address the lack of suitable data about dust source areas and exposure patterns in Africa and to create the possibility of a new generation of dust emission management plans that could improve other parts of the dust cycle such as mitigating derived respiratory diseases.*

References

Braun L, Kisting S (2006) Asbestos-related disease in South Africa: the social production of an invisible epidemic. Am J Public Health 96(8):1386–1396. https://doi.org/10.2105/AJPH.2005.064998. Accessed 13 Apr 2024

Bredeck G, Busch M, Rossi A, Stahlmecke B, Fomba KW, Herrmann H, Schins RPF (2023) Inhalable Saharan dust induces oxidative stress, NLRP3 inflammasome activation, and inflammatory cytokine release. Environ Int 172:107732. https://doi.org/10.1016/j.envint.2023.107732. Accessed 22 Apr 2025

Brouwer DH, Rees D (2020) Can the South African milestones for reducing exposure to respirable crystalline silica and silicosis be achieved and reliably monitored? Front Public Health 8. https://doi.org/10.3389/fpubh.2020.00107. Accessed 26 Jun 2020

Caceres JD, Venkata AN (2023) Asbestos-associated pulmonary disease. Curr Opin Pulm Med 29(2):76–82. https://doi.org/10.1097/MCP.0000000000000939. Accessed 02 Aug 2023

Coleman SRM, Menson KE, Kaminsky DA, Gaalema DE (2023) Smoking cessation interventions for patients with chronic obstructive pulmonary disease: a narrative review with implications for pulmonary rehabilitation. J Cardiopulm Rehabil Prev 43(4):259–269. https://doi.org/10.1097/HCR.0000000000000764. Accessed 23 Aug 2025

Council of the European Union, European Parliament (CEU EP) (2023) Directive (EU) 2023/2668 amending Directive 2009/148/EC on the protection of workers from the risks related to exposure to asbestos at work. Official Journal of the European Union, Series 2023/2668. https://www.europeansources.info/record/proposal-for-a-directive-amending-directive-2009-148-ec-on-the-protection-of-workers-from-the-risks-related-to-exposure-to-asbestos-at-work/. Accessed 16 Oct 2024

Davies TC (2008) Environmental health impacts of East African rift volcanism. Environ Geochem Health 30(4):325–338. Springer Velag, Amsterdam, The Netherlands. https://doi.org/10.1007/s10653-008-9168-7. Accessed 19 Oct 2024

Degrendele C, Prokeš R, Šenk P, Jílková SR, Kohoutek J, Melymuk L, Přibylová P, Dalvie MA, Röösli M, Klánová J, Fuhrimann S (2022) Human exposure to pesticides in dust from two agricultural sites in South Africa. Toxics 10(10):629. https://doi.org/10.3390/toxics10100629. Accessed 19 Oct 2024

Doumbia T, Liousse C, Ouafo-Leumbe M-R, Ndiaye SA, Gardrat E, Galy-Lacaux C, Zouiten C, Yoboué V, Granier C (2023) Source apportionment of ambient particulate matter (PM) in two western African urban sites (Dakar in Senegal and Bamako in Mali). Atmosphere 14:684. https://doi.org/10.3390/atmos14040684. Accessed 19 Oct 2025

Ehrlich R, Akugizibwe P, Siegfried N, et al (2021) The association between silica exposure, silicosis and tuberculosis: a systematic review and meta-analysis. BMC Public Health 21:953. https://doi.org/10.1186/s12889-021-10711-1. Accessed 18.10.2024

Fernandes R, Dupont S, Lamaud E (2023) Experimental evidence of dust flux size distribution variation along two consecutive erosion seasons. Aeolian Res 61(100863). https://doi.org/10.1016/j.aeolia.2023.100863. Accessed 19 Oct 2024

Formenti P, Di Biagio C (2024) Large synthesis of *in situ* field measurements of the size distribution of mineral dust aerosols across their lifecycle, Earth System Science, Data Discuss. [preprint], https://doi.org/10.5194/essd-2023-481. Accessed 19 10.2024), In Review, 2024

Formenti P, Schütz L, Balkanski Y, Desboeufs K, Ebert M, Kandler K, Petzold A, Scheuvens D, Weinbruch S, Zhang D (2011) Recent progress in understanding physical and chemical properties of African and Asian mineral dust. Atmos Chem Phys 11:8231–8256. https://doi.org/10.5194/acp-11-8231-2011. Accessed 19 Oct 2024

Frank AL, van Zandwijk N (2024) Asbestos history and use. Lung Cancer (Amsterdam, Netherlands) 193:107828. https://doi.org/10.1016/j.lungcan.2024.107828. Accessed 08 Sept 2025

AL Frank 2020 Global use of asbestos–legitimate and illegitimate issues Journal of Occupational Medicine and Toxicology (London, England) 15 16 https://doi.org/10.1186/s12995-020-00267-y.Accessed22.04.2025

Fubini, B. and Feneglio, I. (2007). The toxic potential of mineral dust. ELEMENTS 3, 407–414.https://www.elementsmagazine.org/wp-content/uploads/archivearticles/e3_6/fubini.pdf. Accessed 02.08.2023

Fuller, R., Landrigan, P.J., Balakrishnan, K., Bathan, G., Bose-O'Reilly, S., Brauer, M., Caravanos, J, Chiles, T., Cohen, A. et al. (2022). Pollution and health: A progress update. Lancet Planet Health 6 (6): e535–e547. https://doi.org/10.1016/S2542-5196(22)00090-0. Epub 2022 May 18. Erratum in: Lancet Planet Health. 2022 Jun 14. https://www.thelancet.com/journals/lanplh/article/PIIS2542-5196(22)00090-0/fulltext. Accessed 31.07.2023

VE Georgakopoulou C Taskou A Diamanti D Beka P Papalexis N Trakas DA Spandidos 2024 Saharan dust and respiratory health: Understanding the link between airborne particulate matter and chronic lung diseases (Review) Experimental and Ttherapeutic Medicine 28 6 460 https://doi.org/10.3892/etm.2024.12750.Accessed23.08.2025

Gomes, J., Esteves, H. and Rente, L. (2022). Influence of an extreme Saharan Dust Event on the air quality of the West Region of Portugal. Gases 2022, 2 *(3)*, 74–84. https://doi.org/10.3390/gas es2030005. Accessed 19.10.2024

HEI (Health Effects Institute) (2022). The State of Air Quality and Health Impacts in Africa. A Report from the State of Global Air Initiative. Boston, MA: Health Effects Institute. ISSN 2578–6881. https://www.stateofglobalair.org/sites/default/files/documents/2022-10/soga-africa-report.pdf. Accessed 30.06.2023

Hoskin J, Kaneko K, Munakomi S, and Fife, T. (2024). Superficial Siderosis. US National Library of Medicine. (2024). https://www.ncbi.nlm.nih.gov/books/NBK603744/. Accessed 19.10.2024

Hoy RF, Chambers DC (2020) Silica-related diseases in the modern world. Allergy 75:2785–2797. https://doi.org/10.1111/all.14202. Accessed 18 Oct 2024

Huang, X., Rom, X., Gordon, T. and Finkelman, R.B. (2006). Interaction of iron and calcium minerals in coals and their roles in coal dust-induced health and environmental problems. Reviews in Mineralogy and Geochemistry (2006), 64 (1), 153–178. https://doi.org/10.2138/rmg.2006.64.6. Accessed 02.08.2023

Huboyo, H.S., Samadikun, B.P., Astuti, W., Khasanah, N., Ardiyanta, I and Nugroho, A.R. (2025). The impact of industrial air pollution in areas not specifically designated for industrial areas. Air Quality, Atmosphere and Health 18, 675–689 (2025). https://doi.org/10.1007/s11869-024-01668-2. Accessed 18.10.2025

IARC WGECRH (2016) International Agency for research on cancer working group on the evaluation of carcinogenic risks to humans. Outdoor Air Pollution. International Agency for Research on Cancer (IARC), Lyon (FR). (IARC Monographs on the Evaluation of Carcinogenic Risks to Humans, No. 109.) 1.2, Sources of air pollutants. https://www.ncbi.nlm.nih.gov/books/NBK 368029/. Accessed 16 Oct 2024

Klebe S, Leigh J, Henderson DW, Nurminen M (2019). Asbestos, smoking and lung cancer: an update. Int J Environ Res Public Health 17(1):258. https://doi.org/10.3390/ijerph17010258. Accessed 22 Apr 2025

Li T, Cohen AJ, Krzyzanowski M, Zhang C, Gumy S, Mudu P, Pant P, Liu Q, Kan H, Tong S, Chen S, Kang U, Basart S, Touré NE, Al-Hemoud A, et al (2025) Sand and dust storms: A growing global health threat calls for international health studies to support policy action. Lancet Planet Health 9(1):e34–e40. https://doi.org/10.1016/S2542-5196(24)00308-5. Accessed 23 Aug 2025

Maboso B, Te Water Naude J, Rees D, Goodman H, Ehrlich R (2023) Difficulties in distinguishing silicosis and pulmonary tuberculosis in silica-exposed gold miners: a report of four cases. Am J Ind Med 66(4):339–348. https://doi.org/10.1002/ajim.23460. Accessed 18 Oct 2024

Markowitz SB (2022) Lung cancer screening in asbestos-exposed populations. Int J Environ Res Public Health 19(5):2688. https://doi.org/10.3390/ijerph19052688. Accessed 02 Aug 2023

Marques Da Silva V, Benjdir M, Montagne P, Pairon J-C, Lanone S, Andujar P (2022) Pulmonary toxicity of silica linked to its micro- or nanometric particle size and crystal structure: a review. Nanomaterials 12:2392. https://doi.org/10.3390/nano12142392. Accessed 18 Oct 2024

McClellan R (2000) Particle interactions with the respiratory tract. In: Gehr P, Heyder J (eds) Particle–lung interactions. Dekker, New York, pp 3–66

Muteba KM (2018) Mesothelioma incidence and mortality in South Africa from 2003 to 2013. Master's Dissertation, University of the Witwatersrand, Johannesburg, South Africa. https://wiredspace.wits.ac.za/server/api/core/bitstreams/ec07f73a-fa5c-4b51-b95c-57ded07bb197/content#:~:text=The%20official%20statistics%20from%20Stats,2132%20mesothelioma%20deaths%20in%20total. Accessed 18 Oct 2024

Mwangi J (2023) Usage of human diatomaceous earth. Ann Limnol Oceanogr 8(1):009–012. https://doi.org/10.17352/alo.000014. Accessed 18 Oct 2024

Nattrass C, Horwell CJ, Damby DE, Kermanizadeh A, Brown DM, Stone V (2015) The global variability of diatomaceous earth toxicity: a physicochemical and in vitro investigation. J Occup Med Toxicol (London, England) 10:23. https://doi.org/10.1186/s12995-015-0064-7.Accessed 18 Oct 2024

Nelson G (2023) Health effects of silica dust exposure–What do we need to do? Occup Health Southern Africa 29(3). https://hdl.handle.net/10520/ejc-ohsa_v29_n3_a10. Accessed 18 Oct 2024

Nemery B (2022). Metals and the respiratory tract. In: Nordberg GF, Costa M (eds.) Handbook on the toxicology of metals, vol 15th Edn. Chapter 19: General considerations, pp 421–443. https://doi.org/10.1016/B978-0-12-823292-7.00030-9. Accessed 02 Aug 2023

Ofremu GO, Raimi BY, Yusuf SO, Dziwornu BA, Nnabuife SG, Eze AM, Nnajiofor CA (2025) Exploring the relationship between climate change, air pollutants and human health: Impacts, adaptation, and mitigation strategies. Green Energy Resourc 3(2):100074. https://doi.org/10.1016/j.gerr.2024.100074. Accessed 23 Aug 2025

Page MJ, McKenzie JE, Bossuyt PM, Boutron I, Hoffmann TC, Mulrow CD, Shamseer L, Tetzlaff JM, Akl EA, Brennan SE, Chou R, Moher D (2021) The PRISMA 2020 statement: an updated guideline for reporting systematic reviews. BMJ (Clinical Research Ed.) 372:71. 10.1136/bmj.n71. Accessed 16 Oct 2024

Papatheodorou S, Achilleos S (2025) Epidemiological insights into the health impacts of dust storms. Curr Opin Environ Sci Health 46:100626. https://doi.org/10.1016/j.coesh.2025.100626. Accessed 23 Aug 2025

Plumlee GS, Morman SA, Ziegler TL (2006) The toxicological geochemistry of earth materials: an overview of processes and the interdisciplinary methods used to understand them. Rev Mineral Geochem 64:5–57. https://repository.geologyscience.ru/bitstream/handle/123456789/48659/Plum_06.pdf?sequence=1&isAllowed=y. Accessed 08 Sept 2025

Pouri N, Karimi B, Kolivand A, Mirhoseini SH (2024) Ambient dust pollution with all-cause, cardiovascular and respiratory mortality: a systematic review and meta-analysis. Sci Total Environ 912:168945. 10.1016/j.scitotenv.2023.168945. Accessed 22 Apr 2025

Rees D, Phillips JI (2017) The legacy of *in situ* asbestos cement roofs in South Africa. Occup Environ Med 74(Suppl 1):A1-A166. https://www.nioh.ac.za/the-legacy-of-in-situ-asbestos-cement-roofs-in-south-africa/. Accessed 26 Jun 2023

Ross M, Nolan RP, Langer AM, Cooper WC (1993) Health effects of mineral dusts other than asbestos. In: Guthrie, GD, Mossman BT (eds.) Health effects of mineral dusts. Rev Mineral 28:361–407. Conference: Mineralogical Society of America (MSA) conference on health effects of mineral dusts, Nantucket Island, MA (United States). https://doi.org/10.1515/9781501509711-015/html?lang=en. Accessed 19 Oct 2024

Rowan C, D'Souza RR, Zheng X, Crooks J, Hohsfield K, Tong D, Chang HH, Ebelt S (2024) Dust storms and cardiorespiratory emergency department visits in three Southwestern United States: Application of a monitoring-based exposure metric. Environ Res Health ERH 2(3):031003. https://doi.org/10.1088/2752-5309/ad5751. Accessed 16 Oct 2024

Schraufnagel DE (2020) The health effects of ultrafine particles. Exp Mol Med 52:311–317. https://doi.org/10.1038/s12276-020-0403. Accessed 17 Oct 2024

Skuland T, Låg M, Gutleb AC, Brinchmann BC, Serchi T, Øvrevik J, Holme JA, Refsnes M (2020) Pro-inflammatory effects of crystalline- and nano-sized non-crystalline silica particles in a 3D alveolar model. Partic Fibre Toxicol 17(13). https://doi.org/10.1186/s12989-020-00345-3. Accessed 02 Aug 2023

Sun P, Wang B, Zhang H, Xu M, Han L, Zhu B (2024) Predicting coal workers' pneumoconiosis trends: leveraging historical data with the GARCH model in a Chinese Miner Cohort. Medicine (Baltimore) 103(7):e37237. 10.1097/MD.0000000000037237. Accessed 17 Oct 2024

Tessum MW, Anenberg SC, Chafe ZA, Henze DK, Kleiman G, Kheirbek I, Marshall JD, Tessum CW (2022) Sources of ambient PM2.5 exposure in 96 global cities. Atmos Environ (Oxford, England, 1994) 286:119234. https://doi.org/10.1016/j.atmosenv.2022.119234. Accessed 31 Jul 2023

The European Union (EU) (2022) Questions and answers: towards an asbestos-free future. https://ec.europa.eu/commission/presscorner/detail/en/qanda_22_5678. Accessed 16 Oct 2024

Thives LP, Ghisi E, Thives Júnior JJ, Vieira AS (2022) Is asbestos still a problem in the world? A current review. J Environ Manage 319:115716. 10.1016/j.jenvman.2022.115716. Accessed 08 Sept 2025

Tian W, Hu A, Li K, Guo P, Chai Q, Ma R (2024) Dynamic assessment of dust hazard risk in the reconstruction of old industrial buildings: coupling effects of dust distribution and personnel trajectories. Environ Sci Pollut Res Int 31(2):2687–2699. https://doi.org/10.1007/s11356-023-31264-3. Accessed 18 Oct 2025

Tong DQ, Gill TE, Sprigg WA, Van Pelt RS, Baklanov AA, Barker BM, et al (2023) Health and safety effects of airborne soil dust in the Americas and beyond. Rev Geophys 61:e2021RG000763. https://doi.org/10.1029/2021RG000763. Accessed 18 Oct 2024

Tramontano C, De Stefano L, Rea I (2023) Diatom-based nanomedicine for colorectal cancer treatment: new approaches for old challenges. Mar Drugs 21:266. https://doi.org/10.3390/md21050266. Accessed 18 Oct 2024

Verlynde G, Agneessens E, Dargent JL (2018) Pulmonary talcosis due to daily inhalation of talc powder. J Belgian Soc Radiol 102(1):12. https://doi.org/10.5334/jbsr.1384. Accessed 19 Oct 2024

Wang Q, Gu J, Wang X (2020) The impact of Sahara dust on air quality and public health in European countries. Atmos Environ 241:117771. 10.1016/J.ATMOSENV.2020.117771. Accessed 17 Oct 2024

Wang Z, Zhang J, Yang Y, Cao M, Ma J, Li S, Shao H, Du Z (2024) Current status, trends, and predictions in the burden of coal worker's pneumoconiosis in 204 countries and territories from 1990 to 2019. Heliyon 10(19): e37940. 10.1016/j.heliyon.2024.e37940. Accessed 21 Oct 2025

Wei W, Chen Z, Hao D, Liu X, Ni B (2021) Natural diatomite mediated continuous anaerobic sludge digestion: Performance, modelling and mechanisms. J Clean Prod 329:129750. https://doi.org/10.1016/j.jclepro.2021.129750Get. Accessed 18 Oct 2024

The World Meteorological Organisation (WMO) (2023) WMO highlights efforts to tackle sand and dust storms. https://wmo.int/media/news/wmo-highlights-efforts-tackle-sand-and-dust-storms. Accessed 17 Aug 2025

Yi X, He Y, Zhang Y, Luo Q, Deng C, Tang G, Zhang J, Zhou X, Luo H (2023) Current status, trends, and predictions in the burden of silicosis in 204 countries and territories from 1990 to 2019. Front Public Health 11:1216924. https://doi.org/10.3389/fpubh.2023.1216924. Accessed 17 Oct 2024

Yu H, Zahidi I (2023) Environmental hazards posed by mine dust, and monitoring method of mine dust pollution using remote sensing technologies: an overview. Sci Total Environ 864:161135. 10.1016/j.scitotenv.2022.161135. Accessed 21 Apr 2024

Zhang Y, Zhang H, Zhao J, Zhang X, Cao Z, Jiang B (2023) Efficient adsorption of fluoride ions by polyethyleneimine modified diatomite under different conditions in a fixed bed column. Desalin Water Treat 307:153–161. https://doi.org/10.5004/dwt.2023.29906. Accessed 18 10 2024

Zuluaga-Astudillo D, Ruge JC, Camacho-Tauta J, Reyes-Ortiz O, Caicedo-Hormaza B (2023) Diatomaceous soils and advances in geotechnical engineering–Part I. ApplSci 13:549. https://doi.org/10.3390/app13010549. Accessed 02 Aug 2023

Chapter 6
Impact of Climate Change on Geological Processes and Management of the African Coastal Cities: Implications for Population Health and Food Security

T. C. Davies and X. M. Mkhize

Abstract The African coastal zone consists of the east, west, central, southern and Mediterranean coastal areas and comprises many port cities, including the cities of Durban in South Africa, Lagos in Nigeria and Mombasa in Kenya, the cities used in our illustrative case descriptions. Coastal areas in Africa are home to the highest concentration of the Continent's human populations, with increasing burden of urban migrations, and host to diverse industries. It is now well established that all coastal locations are at risk of climate change effects, such as accelerated sea-level rise, increased global temperatures and changing storm frequencies. The impacts of climate change are likely to worsen many geological hazards that African coastal areas already face—increased shoreline erosion and sediment discharge, coastal flooding, and water pollution. But, according to the *Lancet*, "… the impact of changing climate on health is the most serious health threat of the twenty-first century". The main objective of this Chapter is to present a synopsis of our present understanding of the scale, frequencies and trends of some hazardous geometeorological processes impacted by climate change in African coastal cities, and the effect of these processes on population health and food systems. By means of a systematic review based on the PRISMA (Preferred Reporting Items for Systematic Reviews and Meta-Analyses) protocol and backed by long-term documented medical registry data on probable climate induced maladies, an attempt is made to establish and document recent trends in research largely during the last decade on the effects of climate impacted geological processes on human health, and food systems in African coastal cities. Gaps in knowledge (*italicised*) are identified, and directions for future research are proposed. Such an approach, it is hoped, would enable the development of improved policies and interventions for mitigating the health impacts

T. C. Davies (✉) · X. M. Mkhize
Faculty of Applied and Health Sciences, Mangosuthu University of Technology, Umlazi, KwaZulu Natal Province, Republic of South Africa
e-mail: daviestheophilus2025@yahoo.com

© The Author(s), under exclusive license to Springer Nature Switzerland AG 2026
T. C. Davies (ed.), *Recent Advances in Medical Geology Research in Africa: A Decadal View*, SpringerBriefs in Earth System Sciences,
https://doi.org/10.1007/978-3-032-13754-8_6

(of these processes) and their adverse effects on food systems, as well as buttress the evolution of tangible and effective coping strategies of affected populations in African coastal cities.

Keywords Climate change · African coastal cities · Durban · Lagos · Mombasa · Health impacts · Food systems · Addressing the issues

Introduction

A cursory examination of the existing literature reveals the documentation of several geological and geophysical studies, including seismic and stratigraphical analyses of shear-type margins along the African coastal regions (See, e.g., Séranne and Anka 2005; Ceraldi et al. 2016). Results from these studies have been applied with varying degrees of success to a number of environmental problems in Africa's coastal zones. *However, geological processes operating further away from the continental margin, along the Continent's coastlines, have been much less studied or understood, albeit of greater concern from an environmental standpoint* (See: Almar et al. 2023; Ankrah et al. 2023; Dada et al. 2024); *and there is still a dearth of information on important climate change parameters that influence geological processes, geohazards and risks in the coastal cities; and how these processes, in turn, impinge on population health and food security.*

The rise in sea levels continues as the population of Africa's coastal cities soars. Between 2020 and 2030, Africa's 7 largest coastal cities: Lagos, Luanda, Dar es Salaam, Alexandria, Abidjan, Cape Town, and Casablanca are projected to grow by 40% (48 million people to 69 million) compared with the continent's overall anticipated increase of 27 percent (1.34–1.69 billion) (ACSS 2022).

Coastal areas are vulnerable to sea-level rise, potentially leading to erosion, inundation, degradation of public health and loss of infrastructure and habitat. By virtue of their coastal locations, the cities of Durban (South Africa), Lagos (Nigeria) and Mombasa (Kenya), the cities used in our illustrative case descriptions, are particularly vulnerable to sea-level rise and coastal storms. The rates and frequencies of these processes are on the increase; and, accordingly, urban local governments are rightly stepping up their development of climate change adaptation plans. *However, to date, there has been limited success in the application of climate change adaptation policies of African coastal cities, particularly about moving from strategy development to implementation.* This continues to hamper efforts to understand and guide city climate change actions on the Continent. This Chapter helps to address these gaps (*some of which are highlighted through italicisation in the text*) by providing critical insights into the opportunities and challenges experienced, and the solutions found in the process of developing and implementing climate change strategic developments in African coastal cities.

"Coastal risk" and "vulnerability" are terms used to describe the degree to which coastal areas or zones, systems, people, and places are likely to face harm, destruction or degradation resulting from coastal hazards (Beroya-Eitner 2016). Climate model projections, driven by anticipated future greenhouse gas and aerosol emissions, indicate that Earth will continue to warm, with associated increases in sea level and extreme weather events (OCM 2024). The African Continent, in particular has been warming at a slightly faster rate than the global average, at about $+0.3$ °C per decade between 1991 and 2023 (WMO 2024).

The World Health Organisation (WHO) considers climate change to be a fundamental threat to human health (WHO 2023). Climate accentuated geometeorological processes in Africa's coastal zone may affect health through a range of pathways; for example, as a result of increased floods and droughts, increased frequency and intensity of heat waves, increasing allergens, changes in the distribution of vector-borne diseases, water quality issues, air pollution and effects on the risk of disasters and malnutrition.

The impacts of climate change on health and food systems are likely to worsen many of the other problems that African coastal cities already face. Indeed, climate-related illnesses are already on the rise, accounting for more than half of public health events recorded in the region over the past two decades (WHO 2022). *However, the climatopathology of these illnesses is still imperfectly understood* (Botes and McKenzie 2013; Wright et al. 2024). Yet, a broad range of decision makers, including coastal zone planners, relevant government authorities and geoscientists require accurate and well-researched geoscientific information in managing the coastal zone to bring about the obliteration of these diseases. Much of this research should form the basis for "integrated coastal zone management" (ICZM) on a local, national and international level.

The main objective of this Chapter, therefore, is to gain an in-depth understanding of the scale, frequencies, and historical and present trends of climate accentuated erosion and sedimentation processes, and risks of disruption of the natural nutrient cycling in agricultural systems of African coastal cities. A systematic search of the most current bibliography coupled with recent medical record data has enabled the performance of a succinct analysis of empirical evidence of climate change adaptation and mitigation strategies that have been formulated to counter adverse health outcomes and food insecurity in African coastal cities; including the identification of gaps and shortcomings; and to inform policy and practice (Wannewitz et al. 2024).

It is construed that interventions to mitigate the effects of changing patterns of disease and water and food insecurity would benefit from a proper understanding of the extent to which coastal processes are accentuated by climate change. Such understanding should serve as a postulate for the design of alleviative measures for climate-induced health hazards, and definition of farmer adaptation strategies to cope with the ensuing impacts in an integrated manner.

Search Strategy and Selection Criteria

Considerations were made for the credibility of databases included those of MEDLINE, Scopus, Web of Science, Google Scholar, African Journals Online and the Environmental Science Index. Thematic keyword combinations included those related to "geological processes", "climate change", "floods and extreme weather", "health effects" and "agricultural and food systems", enhancing the likelihood of retrieving comprehensive and multidisciplinary literature encapsulated within the selected timeframe (ca. 2015–2025). The search process, data compilation and analyses were done in line with recommendations in the PRISMA (Preferred Reporting Items for Systematic Reviews and Meta-Analyses) protocol (Page et al. 2021) and backed by long-term documented medical registry data on probable climate induced maladies. The focus was on recent studies (2015–2025) that would ensure that the discourse is based on up-to-date and relevant research data published in high impact journals and other reputable publication outlets. This is particularly crucial given the evolving nature of climate change science and, in the provision of valid comparisons. Only articles published in the English language were considered.

Climate Change, Floods and Droughts

Climate change is making heavy intense downpours and rising water temperatures more common. Of all the climate accentuated geometeorological processes impacting population health and food security in Africa, flooding must be considered as one of the most severe (WMO 2024; See section on "Case Description Summaries", this Chapter, to provide a clearer conceptual background). Due to the increase in extreme flood events, combined with compounding and cascading effects from other hazards, as well as underlying societal vulnerabilities, African coastal cities find themselves in a precarious state (UNU 2023).

Coastal Floods

Coastal floods are the result of storm surges that are associated with tropical cyclones and tsunami; consequently, an overflow of ocean water submerges land that is usually dry. The coastal cities of Durban, Lagos and Mombasa are at risk of a combination of both inland and coastal storm surge flooding during extreme climate accentuated geometeorological events, given the existence in these cities of an ideal set of conditions for these processes (See: "Case Description Summaries", this Chapter).

Coastal Droughts

Although droughts can occur naturally, climate change has generally accelerated the hydrological processes to make them set in much more quickly and become more intense, with many consequences. Sadly, as global warming takes hold, without adequate response measures, projections indicate that over the next century drought risk will increase even further across much of the subtropics and mid-latitudes in both hemispheres, concomitant with a regional decline in precipitation (Cook et al. 2018). Drought may lead to increased poverty levels (restricting access to health care), and compromise crop yields and livestock production, with implication on food prices (Kimutai et al. 2025).

The effect of drought can vary depending on the accumulation period of the events with larger risks being associated with more extended drought episodes. *However, the complex and multivariate nature of drought precludes simplistic explanations of cause and effect, making investigations of climate change and drought a challenging task* (Cook et al. 2018).

Health Risks Posed by Floods and Droughts

Floods and droughts can lead to both communicable and non-communicable diseases (NCD's), as well as increase the risk of food insecurity and malnutrition. According to the WHO (2023), climate change and the epidemic of NCDs, are intertwined, with climate change exacerbating drought conditions.

Diseases Caused by Floods

Coastal flooding poses increased risk of various diseases such as waterborne illnesses like cholera and leptospirosis, as well as vector-borne diseases like malaria and dengue fever. These diseases are spread through contaminated water and breeding grounds for mosquitoes, respectively. Additionally, flooding can lead to injuries, chemical hazards, and mental health issues, particularly for displaced populations. The 2022 study by Suhr and Steinert on the epidemiology of floods in Sub-Saharan Africa after an event, pointed to an increased risk of infection with cholera, scabies, taeniasis, Rhodesian sleeping sickness, malaria, alphaviruses and flaviviruses; *and noted that long-term health effects, specifically on mental health, non-communicable diseases and pregnancy, remain understudied.* Floods may also destroy crops and kill livestock, which may lead to longer term effects associated with malnutrition.

Diseases Caused by Droughts

Droughts in coastal Africa can significantly impact health by increasing the risk of malnutrition and infectious diseases outbreaks (Moyo et al. 2023). Droughts can disrupt sanitation systems, concentrate pathogens in water sources, and reduce access to safe drinking water, leading to increased incidence of waterborne diseases such as cholera and other diarrhoeal diseases (Asmall et al. 2021). Droughts can alter vector habitats and increase exposure to vector-borne diseases like malaria and dengue fever. Africa has the highest global cholera burden (Charnley et al. 2022), several drought-prone regions and high levels of inequity. *Despite this knowledge, research on cholera and drought in Africa is lacking* (Charnley et al. 2021, 2022). The reduced availability of clean water and food sources due to drought-induced crop failures and livestock losses are major contributors to these problems. Droughts may also lead to higher levels of asthma and allergies in the affected population (D'Amato et al. 2020). Drought may lead to increasing poverty levels (restricting access to health care), and compromise crop yields and livestock which has implications on production outputs and food prices (Kimutai et al. 2025).

Windstorms

Violent windstorms are becoming more prevalent in African inland towns and cities. *Yet, few studies have reported on their increasing occurrences in Sub-Saharan Africa's low-density and medium-sized cities and towns; but there is the lack of suitable data for the purpose of, amongst others, developing strong wind statistics, disaster models for the built environment or estimations of tornado risk* (Kruger et al. 2016). It is difficult to describe or define a wind hazard holistically, as the adverse effects of strong winds are often exacerbated by the presence of other factors such as dust, low humidity and in strong or severe thunderstorms and heavy precipitation (See, e.g., Nhantumbo et al. 2023). To exemplify the kind of destruction that can be wreaked by severe windstorm in Sub-Saharan Africa, Kafi et al. (2021) used Enhanced Fujita's F-scale matrix to estimate the intensity and rank the damage severity of a windstorm on June 16, 2018 in Bauchi City in northern Nigeria, that destroyed more than 30 lives and thousands of buildings and other essential structures.

Soil Erosion and Food Security

Soil security and food and nutrition security are inextricably linked (Gitton 2024). Soil degradation results in food insecurity and sustainable management is needed for its obviation. In Africa, the picture is quite troubling. The loss of soil from agricultural lands has, for many years, been recognised as a significant problem affecting food production and water quality. Climate change in Africa is expected to reduce food production contributing to increased risk of undernutrition (Moyo et al. 2023; Mehra et al. 2024), as well as food-, vector- and waterborne diseases (Moore and Colwell 2025).

This situation is set to worsen as climate change increases the rate of soil erosion, soil acidification and salinisation, as well as loss of soil organic carbon/organic matter, soil compaction, soil contamination and soil sealing. *Maintaining optimum food production will require the framing of proper land-use policies taking due cognisance of those factors that are degrading the soil ecosystem* (Gebrehiwot 2022).

Soil erosion reduces cropland productivity and contributes to the pollution of adjacent watercourses, wetlands and lakes. *Gully erosion* is the type of soil erosion that consists of an open, incised and unstable channel generally greater than 30 cm deep, and sufficiently large to disrupt normal farming operations by reducing the area of land available for farming. Gully erosion occurs through the removal of the topsoil along drainage channels by surface water runoff.

In Africa, gully erosion most commonly occurs in areas where the contributing effects of land use, climate change and slope interactions are prominent. Gully erosion scours the land, deposits sediments on growing crops, worsens soil quality in the accumulation zones, general drying of the affected regions by interrupting the groundwater table, silting of river channels and pollution of surface waters.

Gully erosion has become a field of growing interest among the research community worldwide; and assessing the impacts of climatic and land use changes on rates of soil erosion by water is the objective of many national and international research projects; *Yet,* there *are still numerous knowledge gaps in our understanding of this complex phenomenon, particularly regarding its spatial patterns, dynamics, and effective mitigation strategies* (See, e.g., Fan et al. 2024).

In the African coastal regions, we need accurate information on the extent to which the magnitudes of the sediment loads transported by rivers influence the functioning of agrosystems, through, for instance, their control of material fluxes, water quality, and the aquatic habitats supported by the river. There is also the need to study precisely how farmers' response to climate change can potentially exacerbate, or ameliorate, the changes in erosion rates expected (See, e.g., Shaffril et al. 2024; Zenda 2024). *Farmer adaptation strategies should be predicated on the nature of the erosion process, which needs to be well understood.* Filling in these knowledge gaps would go a long way towards aiding the process of formulation of guidelines and policies for effective agricultural management of the African coastal zone.

Climate Accentuated Rates of Sedimentation and Water Quality

Excessive levels of suspended stream sediment (turbidity) or a change in sediment distribution resulting from more frequent and intense storms can negatively affect ecosystem health. The impacts from changing levels of erosion and sedimentation threaten fish, invertebrates and aquatic vegetation, in particular (de Villiers et al. 2022). Climate changes, such as more frequent and intense rain events and larger quantities of stormwater runoff can increase erosion and result in greater amounts of sediment washing into rivers, lakes and streams.

In Africa, there have been few direct measurements of erosivity and erodibility. These measurements have often been implemented, using the Universal Soil Loss Equation (USLE) which has its own limitations (e.g., Claessens et al. 2008). Soil erodibility is not a fixed parameter and changes with time. In 2017, Laura et al. determined sedimentation rate, granulometry, total organic carbon (TOC) and altitude, and found these processes to be the main drivers governing pollutant concentrations in sediments. These authors have noted that, according to climate change models, the expected increase of rainfall erosivity will enhance soil erosion, and consequently, the sediment flow to reservoirs, potentially increasing coarse grain fractions and thus potentially diluting pollutants.

More recent methods for determining soil erosivity have used 3-hourly TRMM Multi-satellite Precipitation Analysis (TMPA) precipitation data (e.g., Vrieling et al. 2014), bias-corrected-downscaled (BCD) climate models (Adeyeri et al. 2024) and remote sensing and GIS tools (e.g., Salhi et al. 2023).

Stronger storms, higher river levels, and faster stream velocity can increase erosion and result in increased suspended sediment (turbidity) in water bodies as well as affect normal distribution of sediment along river-, lake- and stream-beds. These climate impacts can challenge efforts to maintain water quality through effective erosion and sediment control management efforts (Firoozi and Firoozi 2024).

Climate change threatens the quality of source water through increased runoff of pollutants and sediment, decreased water availability from drought, and saltwater intrusion (Konko et al. 2024; See also section on: "Coastal Salinisation", this Chapter), as well as adversely affecting overall efforts to maintain water quality. An increase in stormwater runoff can degrade water quality and worsen existing pollution problems (Zhou et al. 2021). Higher air temperatures, and the corresponding increase in water temperatures, can also promote increased growth of algae and microbes in some water bodies. Bacteria and viruses thrive in these new conditions and when they meet humans, can cause numerous illnesses. An increase in Harmful Algal Blooms (HABs) can threaten availability of source water and increase the need for drinking water treatment (Igwaran et al. 2024). There is increasing evidence of the interconnectedness between poor water quality on human health (through water consumption) and agricultural production through the use of irrigation water (Gerdes et al. 2022). This paper underscores how both these realities have negative implications of compromising health status and food security at the household and community levels.

Loss of Wetlands

Wetlands in Africa are comprised of brackish mangrove swamps, forest swamps and marshes. The mangroves are very important ecologically because they provide spawning and nursery grounds for many coastal fish species, and the habitats for many marine invertebrates. Increased salinity of mangroves and freshwater swamps will lead to the destruction of the mangroves not resistant to the highly saline conditions (Naidoo 2023).

The future of Africa's wetlands is closely associated with human wellbeing, such that the effects of climate change cannot be delinked from the human activities that take place in and around wetlands (See, e.g., Mitchell 2013; Khelifa et al. 2022). Widespread poverty creates the need for people to exploit the benefits derived from these ecosystem services and this makes people vulnerable to climate change. This leads to a situation where the threats to wetlands are from a complex mix of factors rather than from a single source.

For example, impoundments trap sediment with associated nutrients starving downstream areas (particularly deltas) of sediment needed to retain their structure and function. This is a problem with Africa's heavily populated deltas such as the Niger and Nile deltas (See: Dada et al. 2015). Several impoundments have been built to supply water to irrigation schemes, often at substantial cost to downstream wetlands. For instance, the fishery of the Logone Valley, which flows into Lake Chad, was reduced by 90% because wetland inundation was stopped through dam construction to service irrigation (Odada et al. 2005; Murumkar et al. 2020).

Waste Disposal into Coastal Environments

Municipal Solid Waste Management (MSWM) in African coastal cities is a known major contributor to climate change as a result of the release of harmful substances during waste treatment and disposal activities. The physical and ecological disturbance caused by indiscriminate dumping often results in widespread physical and/ or chemical changes, too. The natural decomposition of MSW produces carbon dioxide and methane, while improper controls can result in severe climate effects, such as greenhouse gas (GHG) emissions (Liu et al. 2024; Zhang et al. 2024). Landfills emit the most GHG during the waste treatment, compared to anaerobic digestion (AD) (Liu et al. 2024).

Poor waste disposal practices in African coastal cities result in significant health risks due to exposure to contaminated environments and the spread of vector-borne diseases. Inadequate handling of MSW gives rise to a range of other environmental issues, including the contamination of oceans and drains, the occurrence of floods, and the transmission of infectious diseases through the breeding of vectors (Zhang et al. 2024).

Other health issues wrought by improper MSW treatment include risks posed to all living organisms by pollution of air, water and nearby soil (Liu and Hung 2023. Low waste collection rates, open dumping, and burning also contribute to air, water, and soil pollution, impacting public health. Furthermore, the lack of proper waste management can create breeding grounds for disease vectors, increasing the risk of malaria, cholera, and other illnesses.

In extreme cases, as for example with radioactive waste, the threat to health is a direct one. "At very high doses, radiation can impair the functioning of tissues and organs and produce acute effects such as nausea and vomiting, skin redness, hair loss, acute radiation syndrome, local radiation injuries (also known as radiation burns), or even death." (WHO 2023). Demographic growth in Africa, combined with rapid urbanisation, continues to pose a severe challenge in the MSW sector; and so far, existing policies for sustainable waste management are having insufficient impact in the face of ever-increasing waste production and the resulting greenhouse gas emissions (Sama and Bérenger 2023). Imposition of new environmental standards by international decree should lead to much needed improvements.

Navigating the Energy Usage, Climate Change and Health Conundrum

According to Zi et al. (2023), "Coastal cities consume 60% of global energy, and seawater-cooled district cooling systems (SWDCS) and rooftop solar photovoltaic systems (SPVS) are effective technologies for utilizing the abundant renewable resources of seawater and solar energy in coastal regions. …". African coastal cities depend heavily on fossil fuels, whose uses include power production, the desalinisation of water, for transport, tourist activities and the exploitation of marine resources. Some cities even produce and export fossil fuels as a source of foreign exchange. The combustion of fossil fuels is fraught with attendant multiple health issues, including asthma, cancer, heart disease, and premature death, through the air pollution that it causes (Bertrand 2021). Thankfully, however, several African cities are actively exploring and implementing renewable energy solutions, such as solar water heating, wind energy, hydropower and biofuels to address their energy needs and reduce carbon emissions (C40 Cities 2021, 2023).

In 2024, Kelly and Radler examined the effect of energy consumption on climate change in many African countries from 2004 to 2019; and found that "Energy Use" as kg of oil, "Fossil Fuel Energy Consumption", and "Energy Depletion" positively affected climate change. It is evident, then, that African coastal cities possess considerable potential for further development of renewable energy by virtue of their location, an abundance of sunshine that gives rise to wind, and adequate rain and plant life that can be tapped for energy. Future developments could therefore include the increased use of wind, solar and ocean thermal and tidal energy conversion to obviate the effects of fossil fuel dependency. The use of bioethanol from agricultural residues in Africa as a renewable energy source to combat climate change impact is considered to hold significant potential for mitigating climate change and improving public health in the region (See, e.g., Same et al. 2024).

Heatwaves and Wildfires

In Africa, extreme temperatures and wildfires have become issues of concern. Wildfire has long been a natural part of many African landscapes. The impact of wildfires depends largely on its timing, intensity, and frequency. Coastal cities, in particular are facing an increased risk of heat due to climate change and rapid urbanisation. With limited adaptive capacity, heatwaves and wildfires continue to pose havoc to public health and ecosystems (See, e.g., Obe et al. 2023).

The smoke produced by wildfires increases carbon dioxide emissions, significantly reducing air quality, and harming human health in multiple ways. Airborne matter containing very fine particulates and high concentrations of chemicals from forest fires can be particularly harmful since it that can cause respiratory problems and lead to the development of cancers over time.

Chapungu et al. (2024) describe how climate change poses significant risks through temperature and fire regimes on coastal national park-based tourism in South Africa through its effect on accessibility, comfort levels, and spatio-temporal changes of attractions. *Due to the widespread occurrence of fires in national parks along coastal areas of Africa, and the noticeable changes in temperature regimes, more needs to be done to deepen understanding of their trends and impacts.*

Heatwaves

Heatwaves continue to pose significant risks to human health, livelihoods and ecosystems of African coastal cities due to climate change, with more frequent, intense, and longer-lasting heat events. Heatwaves are primarily caused by high-pressure systems that trap warm air in a particular area, preventing it from dissipating. Climate change is a significant factor contributing to the increasing frequency and intensity of *heatwaves.* Other factors contributing to the formation of heatwaves include urbanisation and land-use changes.

The high-pressure systems, also known as *anticyclones,* create a dome of heat by forcing air to sink and compress, which increases temperatures at the surface. The lack of cloud cover under these high-pressure systems allows for greater solar radiation, further heating the ground and the air above it. This combination of factors can result in prolonged periods of excessively high temperatures, often lasting several days to weeks.

Anticyclones can bring extended periods of calm, clear weather and heat to coastal cities, and can potentially impact public health; not directly causing harm in the same way as cyclones, but indirectly doing so, by contributing to problems such as *heat stress, dehydration,* and *exacerbation of respiratory conditions* due to their influence on temperature and air quality.

The highest temperature anomalies in 2023 were recorded across northwestern Africa, especially in Morocco, coastal parts of Mauritania and northwest Algeria (WMO 2024). Several other countries including Mali, the United Republic of Tanzania, and Uganda reported their warmest year on record. More intense and frequent heatwaves and fires are expected to increase the risk of injury, disease, and death. *There is however a general lack of comprehensive heat risk assessment in these areas due to unavailability of urban weather data* (WHO 2023).

Transportation and Communication

Climate change, sea-level rise, and associated increases in climate-related risks pose significant threats to infrastructural elements of transportation in coastal cities (Martello and Whittle 2023). Increasing rise in sea level will result in flooding of transportation corridors and communication networks. Many airports situated on the coastal plains would be flooded, rendering them hazardous for air transportation. This is particularly the case in low-lying coastal cities, such as Durban, Lagos and Mombasa where exposure to natural hazards and disasters such as flooding, storm surges and extreme heatwaves are on the rise [See section on: "Case Description Summaries", this Chapter].

Transportation and communication in cities along the African coastal zone comprise of an extensive network of roads, railways, airports, seaports and canals, which is a reflection of population and industrial growth within the coastal region. There is a vast array of means of transportation within this network, ranging from informal minibus taxis and motorcycle taxis to Bus Rapid Transit (BRT) systems and, in some cities, light rail. Non-motorised transport, like walking and cycling, is also being encouraged to improve accessibility and connectivity.

Grimett (2024) *reviews the impacts of climate change with specific reference to road, rail and coastal transport infrastructure in South Africa and provides insights and recommendations on how coastal transport sector can be protected, moni-tored and upgraded to prevent and limit climate change related damage.* Grimett (2024) *also provides recommendations on possible mitigation measures that would be required to increase the pace of transition towards renewable energy adoption within the transport sector.*

Martello and Whittle (2023) noted that, to improve resilience of the transporta-tion infrastructure under the climate change milieu, it is necessary to understand the inherent system characteristics, and relationships to local and regional socio-economic and socio-political systems. These authors went on further to provide an overview of the theoretical and practical dimensions of the design of climate-resilient transportation systems and relevant dimensions for infrastructure adaptation and planning, including valuation and assessment of equity.

Highlighting existing gaps in literature, it should be noted that further research is needed to better relate natural hazard exposure to physical and operational consequences (e.g., disruption durations, asset-level damages, interdependencies) and improved methods for assessing the adaptive capacity of organisations managing transportation infrastructure systems.

Several institutions from the public and private sectors, such as UN organisations, Multi-National Development Banks (MDBs), non-government organisations (NGOs) and other transport-related entities are forming collaborative mechanisms to mobilise actions on transport and climate change. *However, although political and corporate leadership on transport and climate change is growing in scope and intensity, within and outside of global agreements (See, e.g., SLoCaT 2018), important gaps in knowledge in the climate change and transport milieu remain as at the start of the present decade. These knowledge gaps include the lack of effectiveness in transferring climate adaptation research in practice and the difficulty in implementation. The need for future research to claim adaptive leadership to support timely decisions can never be overstated* (SLoCaT 2018; Arteaga et al. 2023).

The impacts of transport on health are quite significant. For instance, by promoting such physical activities, as walking and cycling, urban transport systems can also influence health outcomes in a positive way (Ducruet et al. 2024). On the contrary, urban transport can create direct, negative impacts on public health through its contributions to local air and noise pollution, as well as road injuries and fatalities. In addition, an effective transport system allows the population to better access health care services (See, e.g., Varela et al. 2019).

Coastal Salinisation

Within the African coastal zone which encapsulates major cities, diverse and growing coastal activities, both domestic and industrial, depend on groundwater sources, because of the absence of large surface water supplies. This has exerted big strains on coastal groundwater resources. The abstraction of groundwater from freshwater coastal aquifers has led to both the landward encroachment of saline waters and the subsidence of coastal lands, rendering them more susceptible to flooding. The island states of Africa, *viz.*, Cape Verde, Comoros, Madagascar, Mauritius, Seychelles, and São Tomé and Príncipe are already experiencing this phenomenon, and the situation is expected to get worse with sea level rise (Vousdoukas et al. 2023; Archer et al. 2024).

The depth of the water table in the coastal zone which is often very shallow, is subject to saline water contamination and pollution (Tully et al. 2019). An increased global sea level rise is expected to raise the water table along the coast even further, and result in increased salinity of the groundwater. Coastal aquifers in cities provide water for drinking and irrigation of crops. Populations consuming saline drinking water are at greater risk of high blood pressure and potentially other adverse health outcomes (Mueller et al. 2024).

Salinity stress consequent upon the more persistent ingress of sea water may also lead to the disruption of coastal fishery causing a disorganisation of the faunal assemblages, thereby resulting in the redistribution of species and failures in the reproduction and survival of their eggs and spores as well as larval/sporophytes.

Studies on groundwater salinisation on the health of adjacent populations in Africa are limited; but valuable insights can be gleaned from studies from other developing regions such as India (See, e.g.: Nayak and Nandimandalam 2023) *which illustrate the associated health risks of salinisation, e.g.,the effect of toxic metal contamination trends derived therefrom.*

Deforestation

The African coastal zone is richly endowed with vast areas of productive mangrove ecosystem. The scale at which deforestation contributes to the loss of biodiversity and climate change in Africa is enormous (Mash 2019) and is well-illustrated in the North African coastal cities (Babapoorkamani and Ricci 2025). The Niger Delta Region, the largest river delta in Africa has much of its area decimated by erosion, flooding and increasing salinisation of groundwater and soil (Sam et al. 2023), all of which processes are exacerbated by climate change. Deforestation in the Niger Delta is expected to worsen with sea level rise (Enaruvbe and Atafo 2016). Plants in deltaic environments not tolerable to increased salinity will simply be decimated.

Deforestation has increased human exposure to mosquito vectors and malaria risk in Africa, but there is little understanding of how socio-economic and ecological factors influence the relationship between deforestation and malaria risk (See: Estifanos et al. 2024).

Impact of Climate Accentuated Geological Processes on Regional Agricultural Systems and Food Security in Africa's Coastal Zone

Agriculture is known to be a significant sector in Africa as a whole. In the coastal zone of Africa, geological processes (flooding, erosion, gullying, sedimentation and so on), exacerbated by global warming are drastically altering the landscape, with dire consequences for agricultural systems and food security, in addition to public health and ecosystem integrity. Changes in erosion and sedimentation rates have been shown to have important implications for soil resource management and sustainable food production (WMO 2024). Many farmers see firsthand, the impacts of extreme weather events, such as storms and heatwaves, as well as slow-moving events like soil erosion, rising temperatures, and changes in water supply—resulting in a decline in agricultural productivity, the nutritional value of crops, food security, and livelihoods.

The processes referred to, lead to degradation of agricultural lands through alteration of the geochemical balance of the soil; and hence, destruction of soil health. Soil health and appropriate nutrient management are critical components of sustainable agriculture, influencing crop yield, environmental sustainability, and overall food security. The increase in extreme weather events predicted with climate change will also magnify the existing water and wind erosion situations and create new areas of concern in the context of agricultural production. For example, in 2023 alone, extreme weather conditions caused widespread floods and below-average rainfall, leading to significant food production and supply shortages in several African countries, including the Central African Republic, Kenya, and Somalia (See: WMO 2024).

The interactions between climate and soil nutrient circulation patterns in Africa are complex and fascinating but remain poorly understood. Understanding the controls and processes that determine the cycling, and the resulting availability of nutrients still remains a key challenge, chiefly because most of the studies focus only on the immediately available pools in soils and in plants. It is important that we increase our knowledge of the rates and magnitudes of erosion and deposition on agricultural lands, since these processes may have a far greater effect on nutrient cycling and availability than is generally believed. Spatially explicit predictions of future erosion rates under changing climatic conditions must be undertaken by studying past and present trends and making dynamic modelling experiments (See, e.g., Teku and Workie 2025). Farmland must be protected as much as possible, with special attention to higher risk situations that leave the soil vulnerable to erosion.

Soil salinisation, exacerbated by climate change, also poses significant threats to agricultural productivity, land restoration, and ecosystem resilience. Tang et al. (2024) review current knowledge on plant-soil interactions as a strategy to mitigate soil salinisation induced by climate change, with a focus on their role in soil salinity dynamics and tolerance mechanisms.

Other significant issues linked to climate accentuation of geological processes in the agricultural milieu include the malnutrition and hunger due to adverse weather on agricultural production, long-term health and development challenges in children, as well as other infectious diseases such as malaria (See under: "Malaria", this Chapter).

Summary Statements on Assessment of the Climate Change—Public Health Nexus in Africa's Coastal Cities

Climate change negatively impacts human health in numerous ways. The effect of coastal development on public health under a climate change scenario can be long-term. Such development can also physically alter the habitat. By reducing permeable surface area, the rate of run-off increases and impacts on water quality by transporting sediments, toxic chemicals, pesticides, herbicides, pathogens, nutrients and other pollutants to local waterways. With the loss of biodiversity, and other damages to the ecosystem, the self-cleansing property of water is reduced (Zhang et al. 2022), with untold consequences for the health of the usually large coastal populations. Climate change also exerts significant strains on health systems, simultaneously increasing demand for health services while impairing the system's ability to respond.

In Africa, a substantial amount of recent research has been documented on the health consequences of climate accentuated geo-meteorological processes operating around coastal localities. Asmall et al. (2021), for instance, working on the adverse effects of droughts in Africa as a result of high surface temperatures and heatwaves, noted an increased prevalence of anaemia, cholera, scabies and dengue fever, and an increased incidence in child disabilities. Salvador et al. (2024) provide robust evidence of an increased risk of all-cause mortality and specific causes (infectious and parasitic; endocrine, nutritional, and metabolic; circulatory; respiratory) associated with drought events in South Africa. *Despite these recent studies, there remains a dearth of information on health and climate change in the coastal zone of Africa; and more research is therefore urgently required on this subject. In the same vein, the 2023 Lancet Countdown Report concluded that there has been little progress in protecting individuals from the adverse health effects of climate change* (Romanello et al. 2023; see also: earlier report in: Costello et al. 2009). It is clearly evident that global action against climate change needs to move more quickly, as the inequalities in the effects of climate change, including the impact on health and food security are increasing (See: Parums 2024).

Vector-Borne Diseases

As climate change alters temperatures and weather patterns across the world, the potential incidence, seasonal transmission, and geographic range of various vector-borne diseases will increase (Obame-Nkoghe et al. 2024). These diseases include malaria, dengue fever, and yellow fever (all mosquito-borne), various types of viral encephalitis, schistosomiasis (water-snails), leishmaniasis (sand-flies), Lyme disease (ticks), and onchocerciasis (West African river blindness spread by black flies). The formal modelling of the effects of climate change on vector-borne diseases has focussed on malaria and dengue fever. This is so, because the burden of these diseases is huge and are sensitivity to temperature and rainfall changes, which are projected to increase their geographical range and intensity of transmission (Githeko et al. 2000). Both statistical and biologically based (mathematical) models have been used to assess how a specified change in temperature and rainfall pattern would affect the potential for transmission of these and other vector-borne diseases (See, e.g., Colón-González et al. 2021; Brown et al. 2024; Singh and Saran 2024).

Diseases such as Lyme disease, chikungunya, and Zika also show climate-related impacts. *However, malaria and dengue fever provide excellent examples for research on unravelment of the complex interactions involved in their public health impacts.*

Malaria

The pathogen Malaria in humans is an acute or subacute infectious disease caused by parasites of one of six species of the genus Plasmodium (Haemosporida: *Plasmodiidae*) of the phylum Apicomplexa: *P. falciparum, P. vivax, P. ovale wallikeri, P. ovale curtisi, P. malariae*, and *P. knowlesi. M*alaria can also be classified as a water vector (waterborne) disease, since it is transmitted by a vector, mosquito, which needs water or moisture to breed.

In 2022, Mafwele and Lee used malaria incidence data, temperature data and rainfall data collected in 1901–2015 in African regions to construct and analyse climate networks to show how climate relates to the transmission of malaria. Their studies revealed a positive correlation between malaria networks with temperature and rainfall networks. However, *despite the research efforts of various other groups on estimates of how climate change will enhance malaria risk to human health, as of today (2022), we still do not fully understand the impact of climate change on malaria incidence and transmission; and further multidisciplinary research on this subject is required* (See: Megersa and Luo 2025).

Waterborne Diseases

Climate change has significantly influenced the spread of waterborne diseases (WBDs), which affect environmental quality and human life. A WHO analysis of 2022 found that waterborne diseases accounted for 40% of the climate-related illnesses over the past two decades. Cholera and typhoid are examples of bacterial foodborne or waterborne infectious diseases, which are clearly associated with temperature increases (Jung et al. 2023). In Africa, diarrhoeal diseases are the third leading cause of disease and death in under 5 children (Reiner et al. 2018). Yet, it is known that a significant proportion of these deaths is preventable through safe drinking water, adequate sanitation and hygiene. The WHO 2022 analysis also showed that vector-borne diseases, notably yellow fever, accounted for 28% of the climate-related health emergencies, while zoonotic diseases, specifically Congo-Crimean haemorrhagic fever, were the third most prevalent.

Asthma

Asthma is a prevalent chronic respiratory disorder, often resulting in recurrent attacks of wheezing, breathlessness, chest tightness, coughing, and excess mucus. Asthma patients experience worse symptoms with the effects of climate change, such as air pollution and pollen production (Eguiluz-Gracia et al. 2020; Goshua et al. 2023).

Concerns about air pollution in the City of Durban are at a high level, particularly in the industrial areas (C40 Cities 2022). Air pollution levels are particularly harmful in the industrial areas, and especially the South Durban Basin, where the $PM_{2.5}$ annual mean concentration levels can reach 29 $\mu g/m^3$ [almost six times as high as the updated WHO threshold of 5 $\mu g/m^3$ (Pai et al. 2022)]. Such high air pollution levels account for relatively greater incidence rates of asthma and respiratory problems. Indeed, studies by Cities-40 (2022) have shown that children living near industrial areas experience higher rates of asthma and asthmatic symptoms compared to those living further away.

The relationship between air pollution and climate change for the City of Lagos was reviewed by Komolafe et al. (2014); whereas a more recent review by Yussuf et al. (2023) covered climate induced air pollution and paediatric respiratory infections in children in Mombasa and other major cities in Kenya. Recent data collected by Njeru et al. in 2023 using a "Calibrated Network of Low-Cost $PM_{2.5}$ Monitors" in Mombasa (Njeru et al. 2024) indicate polluted air slightly above the WHO health-based guideline value of 5 $\mu g/m^3$ as given in Pai et al. (2022); but substantially lower than other major African cities; and proposed the establishment of long-term open access air quality monitoring stations. Air pollution in Mombasa, particularly high levels of $PM_{2.5}$, is linked to increased risks of acute respiratory infections (ARIs) in children, with long-term exposure potentially exacerbating respiratory problems (Yussuf et al. 2023). *Despite its prevalence, the specific effects of these factors on asthma remain unclear.*

Wang et al. (2024) performed a systematic assessment of the epidemiological evidence using spatial and temporal methods on the impact of climate and environmental factors on childhood asthma. *Coming out clearly in these authors' conclusion, was the need for future research that would integrate spatial and temporal methods to better understand the effects of environmental and climate changes on childhood asthma.*

Case Description Summaries

1. **Durban, South Africa**

Durban is the largest port city on Africa's east coast and the third largest of South Africa's metropolitan areas. Two thirds of the municipality is rural, but urbanisation is progressing rapidly (Roberts et al. 2016; Fig. 6.1). By virtue of its location (eastern coast of South Africa), Durban is vulnerable to a range of climate change impacts, including sea-level rise, increased temperatures, intense storms, flooding and coastal erosion (eThekwini Municipality 2011a). As is the case with other African coastal cities, these phenomena pose serious threats to public health, water resources, food systems and other aspects of the well-being of the City's population, more particularly, the urban poor (Fig. 6.1).

The City is projected to experience higher average temperatures, potentially leading to increased heat stress and water scarcity. More frequent and intense rainfall events are projected; and these will result in increased flood risks and damage to infrastructure (eThekwini Municipality 2022a). Changes in rainfall patterns and increased evaporation rates could lead to decreased water availability, impacting agriculture, sanitation, and overall water security. Durban's coastal aquifers are an important water supply source, especially in the Maputaland area; but face challenges like potential pollution and the risk of saline intrusion from the Indian Ocean.

Of all the impacts of climate change on the City of Durban, the death and destruction caused by flooding must count as the most severe. The frequency of devastating floods continues to increase. Poor infrastructure, inadequate urban planning and the loss of natural buffers amplify the impact of flood events (eThekwini Municipality 2022a).

Climate change can exacerbate existing health problems and create new ones. The latest reports recount the numerous ways in which climate change is expected to negatively impact human health in Durban (See e.g., eThekwini Municipality (2022b)). The exacerbation of existing health challenges, such as the spread of vector-borne and waterborne diseases and increase in heat-related and respiratory illnesses remain of especial concern. Heat waves (three or more consecutive days with maximum temperatures above 30 °C) could increase by about 30%, while extreme heat waves (three or more days above 35 °C) are expected to double in the immediate future (eThekwini Municipality 2011b). More intense and frequent heatwaves and fires are expected to increase the risk of injury, disease, and death.

The risk of malaria and cholera will increase because of warmer temperatures and more frequent flooding (eThekwini Municipality 2011b). The impacts of these events are felt differently by women and men, with the worst impacts falling disproportionately on women, especially those in women-headed households (eThekwini Municipality 2022b).

Durban is a "city of rivers". Several State and Community Development Projects have been put in place to rehabilitate rivers and catchments (Martel and Sutherland 2019; Martel et al. 2022) to protect the city from flooding and improve the quality of life of citizens who live alongside rivers and streams that form part of Durban's 18 major river systems (See section on: "Closing the Gaps: Climate Change Interventions in Combatting Disease and Food Insecurity in Durban, Lagos and Mombasa", this Chapter).

Despite these interventions, residents in the Greater Durban Metro Area (GDMA) continue to suffer untold havoc from frequent flood events. The immediate or short-term impacts of the recent flood events include loss of life, destruction of housing and water and sanitation services, the reduction or complete loss of mobility, loss of income and ability to go to work (Sutherland 2024).

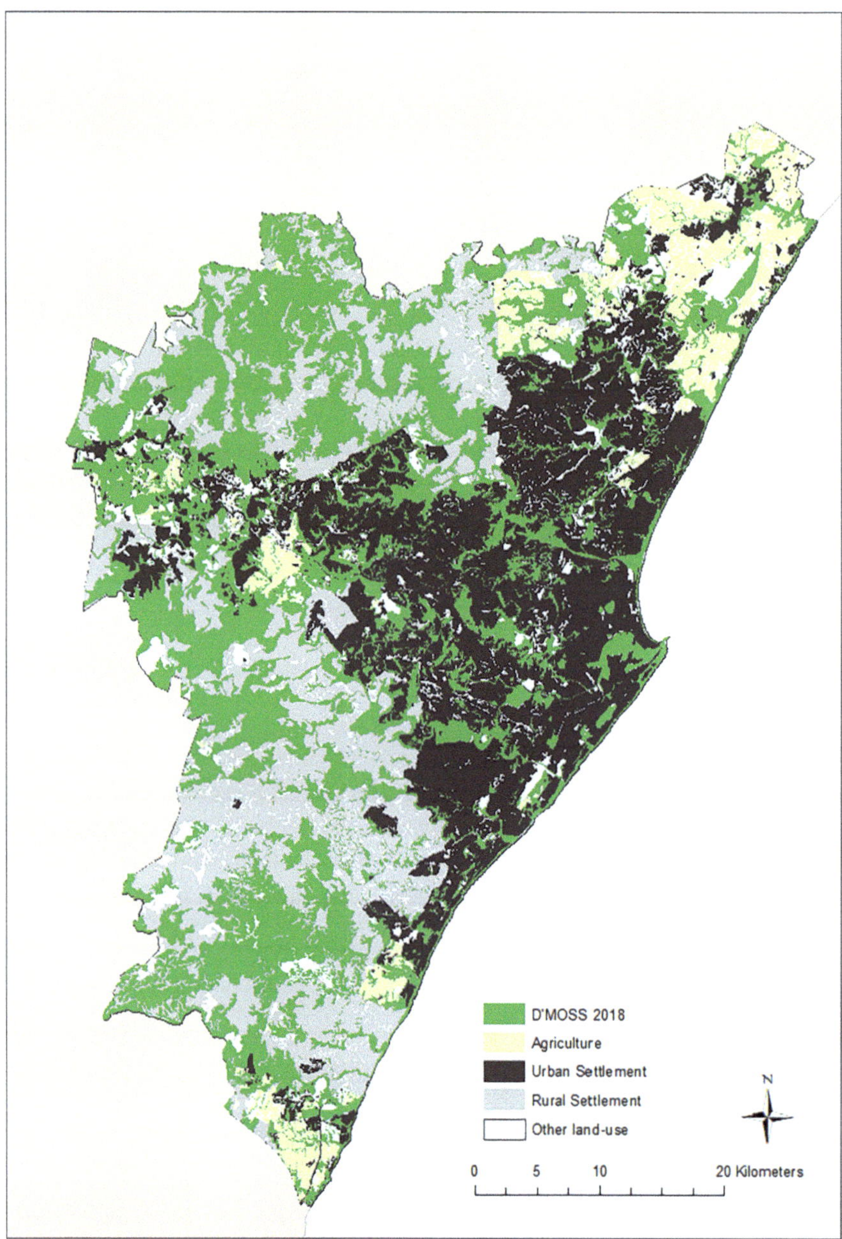

Fig. 6.1 The Durban Metropolitan open space system adjoining the urban/peri-urban areas, all within the eThekwini Municipality. *Credit* Cameron McLean, Biodiversity Management Department, eThekwini Municipality. Reproduced with permission granted on 21.08.2025, Cameron McLean, Senior Specialist Ecologist, Biodiversity Management Department, eThekwini Municipality

Food Security

The City of Durban already faces significant food security challenges through increased temperatures, altered rainfall patterns, and more frequent extreme weather events, impacting crop yields, water availability, and overall agricultural productivity (eThekwini Municipality 2022a). Changes in rainfall patterns and increased temperatures can negatively impact agricultural production even further, potentially leading to food shortages and economic hardship. Reduced food production will contribute to increased risk of undernutrition.

Barring continued research on strategic interventions as advocated in this Chapter, climate change is projected to reduce food production and accessibility even further, contributing to increased risk of undernutrition, particularly for vulnerable populations. Thankfully, the City is working to adapt to these changes and mitigate their contribution to the climate change afflictions [eThekwini Municipality (2022a); See also, under: "Closing the Gaps: Climate Change Interventions in Combatting Disease and Food Insecurity in Durban, Lagos and Mombasa", this Chapter].

2. **Lagos, Nigeria**

Lagos (Fig. 6.2), a coastal megacity in Nigeria, faces substantial climate change challenges, including sea level rise, extreme temperatures, high rainfall intensity, flooding, and other extreme weather events, all of which are exacerbated by the City's low-lying topography, high population density, and inadequate infrastructure. Ndimele et al. (2024) describe life in the City of Lagos as daunting, in the context of climate induced hazards, because the intensity of frequent flood events destroys livelihoods, harms human health, and even causes death in extreme cases.

Human health impacts are brought about through the contamination of water bodies and the increasing incidences of waterborne diseases such as malaria, cholera, typhoid, yellow fever, diarrhoea, leptospirosis and hepatitis A (Echendu 2020). In addition to the immense destruction done to the public health landscape, the annual flooding in Lagos results in impacts on biodiversity and local ecosystems as well (Roa et al. 2022). Ndimele et al. (2024) advocate for the adoption of advanced flood risk management strategies to help in flood containment and management in the City.

The prolificity of research papers, reports and Government Policy Documents during the last decade, on the growing adverse impacts of climate accentuated geo-meteorological processes on population health and food security in Lagos City bespeaks the magnitude of the problem (See, e.g.: Oni et al. 2021; Okeke 2022; Tajudeen et al. 2022; Ndimele et al. 2024; and references within).

It is gratifying to note that in the last two decades, increased political will, the stepping up of local action through popular participation, and development of institutionalised adaptation frameworks have gone a long way in addressing some of the numerous challenges (See: Elias and Omojola 2015; section on: "Closing the Gaps: Climate Change Interventions in Combatting Disease and Food Insecurity in Durban, Lagos and Mombasa", this Chapter).

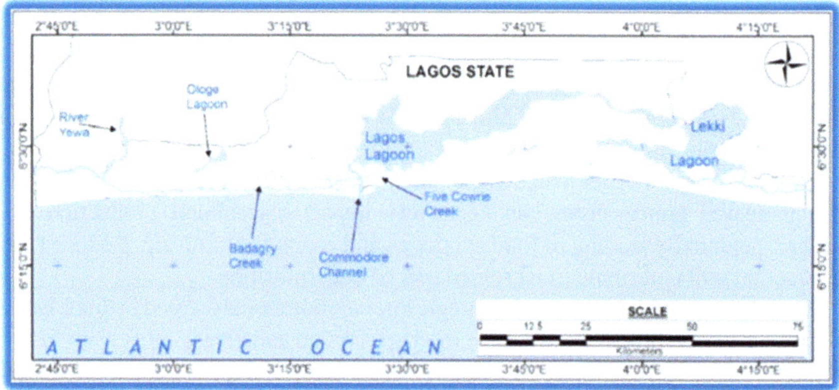

Fig. 6.2 Map showing the Lagos coastal area. *Source* Nwilo et al. (2020). Permission for reuse was granted on 22.08.2025 by Chukwuma John Okolie, co-author who downloaded the image Additionally, the Journal "Geomatics" is an "MDPI Open Access Journal", and states: **Permissions** "No special permission is required to reuse all or part of article published by MDPI, including figures and tables. For articles published under an open access Creative Common CC BY license, any part of the article may be reused without permission provided that the original article is clearly cited. Reuse of an article does not imply endorsement by the authors or MDPI. Furthermore, no special permission is required for authors to submit their work to external repositories. This policy extends to all versions of a paper: submitted, accepted, and published" https://www.mdpi.com/ope naccess (Accessed 22.08.2025)

3. **Mombasa, Kenya**

Mombasa is Kenya's second largest city with a population of over 1 million, according to the 2019 census (KNBS 2019). It is the largest seaport in East Africa, serving Kenya together with many landlocked countries and the north of Tanzania. The city has a long history of disasters reflecting the adverse impacts of climate change, including floods (Fig. 6.3), which cause serious damage nearly every year and, often, loss of life.

Relatively recent accounts of direct and indirect impacts of climate modulated geometeorological processes on population health in the City of Mombasa include the works of Awuor et al. (2008) and Okaka and Odhiambo (2019). More recent accounts are found in the Mombasa County Climate Change Action Plan of 2020–2024 (DEWME 2021) which demonstrates the numerous ways in which climate change stressors influence human health and disease in the city. In 2021, the International Federation of the Red Cross (IFRC/Climate Centre) pointed out the urgent risks of the climate crisis to human health in Kenya as the geographic expansion of climate sensitive vector-borne diseases, an increase in waterborne diseases, and the nutrition implications of food shortages due to longer drier spells, increased land-surface temperature and water scarcity impacting agriculture.

As a result of Mombasa's unique urban geography, ecological risks are concentrated in the *Geology/Water category* (expert weighted average 5.93) of the Stimson Center's Climate and Ocean Risk Vulnerability Index (CORVI), a decision support

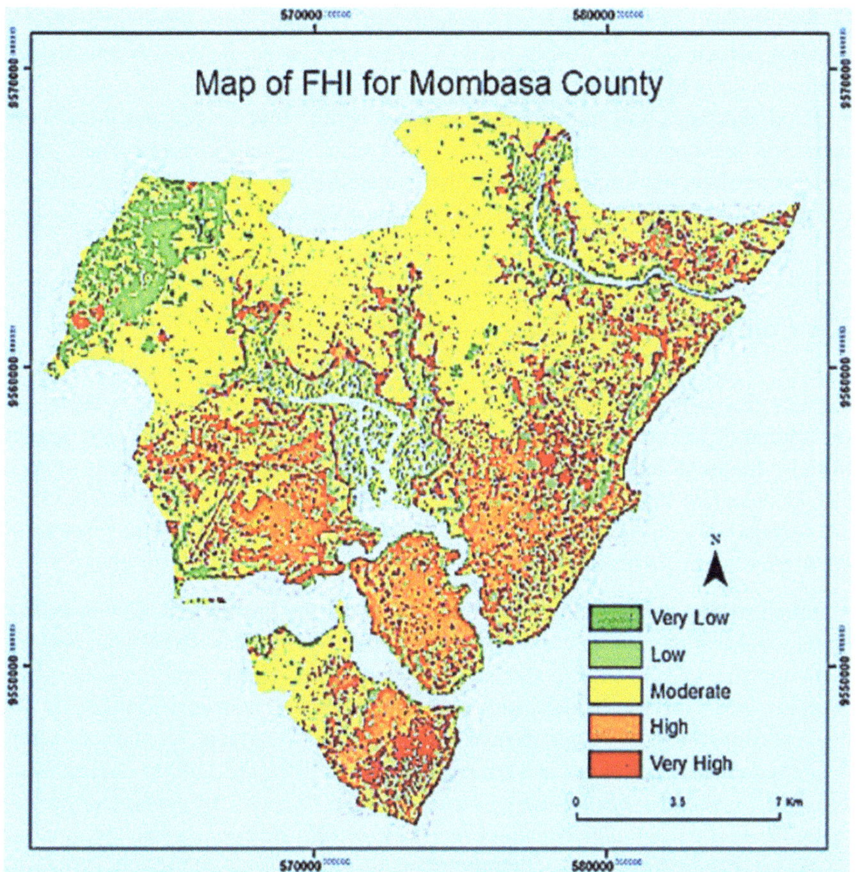

Fig. 6.3 Map of flood hazard index (FHI) for Mombasa. *Source* Hategekimana et al. (2018). "Springer Nature grants permission for authors, readers and third parties to reproduce material from its journals, books and online products as part of another publication or entity. This includes, for example, the use of a figure in a presentation, the posting of an abstract on a web site, or the reproduction of a full article within another journal. …". Reprints & Permissions | Nature Portfolio (Accessed 05.07.2025)

tool which compares a diverse range of ecological, financial, and political risks across 10 categories and nearly 100 indicators to produce a holistic coastal city risk profile (Stuart et al. 2021). This rating system demonstrates the threat posed by ocean risks to the urban centre. Saltwater intrusion in estuaries and freshwater aquifers (expert weighted average 5.94 in the CORVI scale) is another negative consequence of sea-level rise in the City of Mombasa.

The most urgent risk is the increasing heat stress due to rising temperatures. Most recently, Brimicombe et al. (2025) provided insights on the relationship between heat metrics and adverse birth and maternal health outcomes in Mombasa.

Waterborne diseases such as cholera and typhoid are exacerbated through flooding, which also has the potential to expand the range of vector-borne diseases such as malaria by altering mosquito breeding conditions.

Food shortages and nutrition implications result from longer and drier spells, increased land-surface temperatures and water scarcity, impacting agriculture. All of these conditions work toward significantly degrading the local population's health, food systems and general wellbeing of Mombasa residents.

Key Considerations in Addressing the Issues

Climate change is exacerbating health risks in the African coastal zone by increasing the spread of infectious diseases, impacting water quality, and potentially leading to more frequent and severe weather events like floods and storms. These changes can increase the burden of existing diseases and disrupt healthcare access. There are certain key considerations to consider in addressing the issues, in formulating countermeasures, in proffering coping strategies and other interventions.

- Much of the research on coastal change forms the basis for ICZM on a local, national and international level. Shoreline Management Plans (SMPs) are required to consider coastal change and issues such as sediment transport in the nearshore zone. Most SMPs consider distinct parts of the coast, such as complete estuaries or sections of coast in which near-shore sediment is largely "contained" within a coastal cell, or behaves in a consistent manner. *With the African coastal zones, however, because of the lack of local analyses, the scale of geological hazards and risks posed by climate change has yet to be fully determined and documented. This is a prerequisite for addressing the ensuing public health issues.*
- Geological, archaeological and historical records are used to establish the nature of past coastal change. Monitoring of coastal change can also be undertaken by using a broad range of techniques including airborne laser ranging technology (LIDAR) and digital aerial photogrammetry. These techniques are used to determine coastal topography, coastal erosion, and shoreline position with high accuracy. The bathymetry of offshore areas is determined by several geophysical techniques including side-scan sonar or multi-beam surveys.
- Several approaches have been used to prevent gully initiation and reduce gully erosion once started. These have generally been used in combination. The design, implementation and monitoring of an erosion control and sediment production plan involves a significant amount of expertise and resources if the plan is to be successful. It is important that the designer and contractor recognise that successfully implementing erosion and sediment control (ESC) measures requires a good understanding of the principles of the ESC process by both design and field staff. Installing BMPs (Best Management Practices) correctly to specific site conditions and ongoing timely upgrading and maintenance are essential for a successful outcome. Large reductions in elevated sediment loads at the catchment scale

will never be achieved unless gully erosion is addressed cumulatively through innovative proactive land management actions.

- The erection of flood defences (engineered structures or systems such as flood barriers, flood doors, sandbags, and coastal defences, among others), designed to protect areas from the harmful effects of flooding by controlling, redirecting, or preventing the flow of floodwater has often been heralded as an ideal solution. A number of flood-affected European countries have appraised the efficiency of these defence systems (See: Marijnissen et al. 2021; Office of National Statistics (U.K.) 2023) and lessons learnt can be adaptable for application to the African coastal zone. However, several deficiencies of flood defence mechanisms have also been documented (See, e.g., Wang et al. 2022).

- Whatever approach is used, no section of coast should be studied or managed in isolation. The whole picture must be understood, regarding changes in the past, the present position and how any coastal management scheme will be affected by future changes. The best and most sustainable options probably lie in an ICZM approach. This may engender multiple response strategies that can be modified for different socio-economic factors and environmental conditions, working with natural processes rather than against them.

- While careful assessment of coastal changes must form the cornerstone of effective coastal management, it is often insufficient to monitor sequential variations without paying due regard to the causes of change. Knowledge of these causes would obviously assist attempts at stabilisation.

- Depending on future emission trajectories, it may not be possible to prevent the contamination of groundwater drinking resources in high-risk areas, but adequate treatment of extracted water and/or provision of clean drinking water would reduce the health risks to resident populations.

- Soils play a critical role in carbon sequestration and climate regulation (Gitton 2024). Their degradation should, at all costs, be avoided, as this reduces their carbon storage capacity, contributing to global warming.

- Soil degradation impacts several key areas, including agriculture and food security, ecosystems and biodiversity, human health and water quality and climate change resilience (Gitton 2024). The need to support farmers in adopting soil-friendly practices, such as reduced tillage and cover crops, which help reduce erosion and improve soil fertility can never be overemphasised. Scientific evidence garnered so far indicates that climate change is likely to have a significant impact on crop nutrient cycling, uptake and allocation.

- *To be able to predict future effects of climate change on micronutrient migration and distribution, we need to progress from studying effects of single factors to analysing interactions between multiple factors such as elevated CO_2 and temperature or water availability. In Africa, these studies, at the moment, are very limited. We do not as yet possess sufficient breadth of information relating to micronutrient circulation and bioavailability so as to be able to make reliable forecasts and to drive agricultural modernisation.* How soil organic matter levels react to changes in the carbon and nitrogen cycles will influence the ability of soils to support crop growth, which has significant ramifications for food security.

- The harnessing of coastal energy from waves and tides and the siting of conventional or nuclear power plants by the coast give cause for concern. Nevertheless, prospects for exploitation of coastal energy in the African coast are alive. The probable environmental impact of tidal barrages is being investigated (Khojasteh et al. 2022). Broader environmental impacts, as well as techno-economic assessments, are difficult to predict and long-term management decisions associated with harnessing the potential of tidal energy schemes are still awaited (See, e.g., Khojasteh et al. 2022); and major tidal energy projects are yet to be undertaken.

- Sufficient evidence now exists to point to climate, especially heavy rainfall and high temperatures, as having the potential to increase the risk of waterborne diseases (Levy et al. 2018). *Our research efforts should now turn to identifying how and where to intervene to reduce risk in the most vulnerable populations.*

- Approaches to the health management of extreme weather events involve improved early warning systems (EWS), effective contingency planning, and identification of the most vulnerable and exposed communities (de Perez et al. 2022). They include forecasting, predicting possible health outcomes, triggering effective and timely response plans, targeting vulnerable populations, and communicating prevention responses. However, forecasting climate and disease is fraught with several complexities and limitations. Examples of these are as follows: (i) The difficulty of accurately modelling the interplay of various factors such as climate, human behaviour, and pathogen dynamics, (ii) the limitations of current climate models, and (iii) the challenges of translating scientific findings into practical interventions.

- Public health authorities need to upgrade existing emergency programmes and conduct exercises to enhance preparedness for anticipated health risks due to new extreme events such as saline water intrusion into drinking water courses and severe flooding from storm surges. These efforts to improve disaster preparedness must also run in parallel with efforts to strengthen local community resilience.

- Adaptation to indirect effects poses difficult challenges to policy making due to complex causal chains and limited predictability (See, e.g., Wannewitz et al. 2024). A better understanding of the nature of these processes would no doubt enhance our predictive capacity as well as our capacity for upgrading early warning systems for extreme weather events and other geological processes. These indirect impacts still pose serious obstacles for climate adaptation, especially where health responses require integrated and cross-sectoral interventions (Watts et al. 2015).

- According to UNU (2023) coastal city populations recovering from public health emergencies due to climate change need stakeholder and government support with robust short-and-long-term recovery plans throughout the recovery process. Evacuation shelters need to be upgraded, and current policy and strategy documents that outline recovery plans should be reviewed and updated regularly.

- The strengthening of hazard mapping, monitoring, and early warning systems should constitute key elements of long-term action plans. Such plans must prioritise designing resilient buildings, infrastructure and neighbourhoods equipped to live with water and adapt to future conditions.

The Africa Centre for Strategic Studies (ACSS) (2022) proposes the adoption of layered and mutually reinforcing policy responses that must be focussed on fundamental questions about whom cities are for and whom they prioritize as they grow and adapt; and that: "By balancing ecological, economic, and political strategies, municipal and national leaders can strengthen the resiliency of Africa's coastal cities as they adapt to the threat posed by rising sea levels".

Closing the Gaps: Climate Change Interventions in Combatting Disease and Food Insecurity in Durban, Lagos and Mombasa

Knowledge gaps regarding waterborne and vector-borne diseases exist in the coastal areas of Durban, Lagos and Mombasa, including an understanding of the true burden of these diseases, and the effectiveness of interventions, especially in the most vulnerable populations.

Durban, South Africa

The Durban Municipality has been preparing for climate change for several years. Concentrated municipal efforts at combatting climate change effects gathered pace in the early 2000s with an externally driven focus on mitigation (Roberts et al. 2016). As local understanding of climate change improved, Durban emerged as an early leader in the field of climate change adaptation. Notable progress has been recorded in research and policy development as well as the implementation of innovative programmes, particularly in the water sector (See: e-Thekwini Municipality 2022a), and in combatting the incidence of vector-borne and waterborne diseases. Recent developments in implementing policies in the White Paper: "Durban Climate Change Strategy" (DCCS) are articulated in the work of O'Donoghue et al. (2022).

Durban is a member of the C40 Cities network, a global network of mayors taking urgent climate action. The C40 Cities network is committed to taking urgent climate action, aimed at reducing emissions by 40% by 2030 and 80% by 2050, and becoming carbon neutral by 2050. The City has developed a climate intervention roadmap known as the "Durban Climate Action Plan" (DCAP) (See: eThekwini Municipality 2019), part of the broader DCCS (O'Donoghue et al. 2022). The aim of this Plan is to: "… achieve a 40% reduction in emissions by 2030 and an 80% reduction by 2050, with a vision of a sustainable, climate-resilient city prioritising peoples' needs." [Sic]. The plan also incorporates strategies for reducing energy consumption, improving waste management, and enhancing the city's resilience to climate impacts. The plan also takes cognisance of the need to address social equity and the vulnerability of marginalised communities. *However, despite these efforts, gaps remain in our understanding of the geographical spread and mechanism of transmission of diseases influenced by climate change.*

eThekwini Municipality also has several existing programmes to support food access and address food insecurity in Durban, focussing on initiatives like community gardens, urban agriculture, and partnerships with local businesses to ensure sustainable food production and access (Mkhize and Sibanda 2022; eThekwini Municipality 2022b). *To address the anticipated increase in food insecurity in Durban due to climate change, and ensure sustainable food production and access, the Municipality must extend and expand the existing "agriculture and food security programmes", incorporating initiatives such as community gardens, urban agriculture, and partnerships with local businesses* (See: Mkhize and Sibanda 2022; eThekwini Municipality 2022b).

Lagos, Nigeria

In the coastal City of Lagos, as already stated, rising temperatures, flooding and water scarcity have been causing a tremendous increase in the occurrence of waterborne diseases and water pollution. *Earlier efforts at addressing these challenges, according to* Elias and Omojola (2015), *have been haphazard, largely top-down, uncoordinated and fragmented.* This has engendered an urgency in intensification of efforts to protect the health, lives and livelihoods of the people of Lagos, from these devastating effects. These efforts are encapsulated in the recently released reports: "Lagos Economic Outlook: Charting a Resilient Path Towards a Sustainable Future" (LSG 2025) and "Lagos State Climate Adaptation and Resilience Plan (LCARP)" (NESG 2023; MEWR 2024); and recapitulated in the following "paraphrases" from the World Bank and the Federal Government of Nigeria, respectively:

> To address climate change's adverse effects on food security in Lagos, focus has to be on sustainable agriculture, resilient farming practices, and diversification of food systems. This includes promoting climate-smart agriculture, investing in water management, and supporting local farmers with resources and knowledge (World Bank Group 2022).

> To address the adverse effects of climate change on health in Lagos, a multi-faceted approach is needed, including strengthening health systems, promoting climate-resilient infrastructure, and addressing the root causes of climate change, such as air pollution and poor sanitation (FGN 2024).

Mombasa, Kenya

To address the adverse effects of climate change in the health sector, Mombasa County endeavours to "… strengthen health systems, promote climate-resilient infrastructure, and enhance public health awareness and education, while also advocating for policy changes and international cooperation" (DEWME 2021). Other mandates enumerated by the Municipality to address the adverse effects of climate change in the health sector are:

1. Integrate climate change considerations into the health sector.
2. Promote climate vulnerability and risk assessment tools and decision support systems to enhance informed decision making in the health sector.
3. Promote disease surveillance, monitoring and early warning system.
4. Strengthen and promote emergency management services to better handle emergency and disaster situations related to climate change and health.

To address food insecurity stemming from climate change in Mombasa County, the Municipality intends to focus on enhancing climate resilience in agriculture, promoting diversified livelihoods, and strengthening disaster preparedness and response mechanisms, while also supporting sustainable resource management and community engagement" (DEWME 2021).

Conclusion

The combination of accelerated sea-level rise, increased temperatures, shoreline erosion, changing storm frequencies, coastal flooding, water pollution, droughts and loss of wetlands, means that African coastal cities such as Durban, Lagos and Mombasa, having high population density, will be particularly vulnerable to the effects of climate change on health and food security. A broad range of decision makers, including coastal zone planners, relevant government authorities and geoscientists, require accurate and well-researched information in managing the coastal zone.

The dynamics and patterns of erosion and sedimentation in Africa's coastal cities are unique, due to a peculiar nexus of changing micro-climatic conditions, geological and hydrological processes. There is little doubt that the rates of these processes will escalate with enhanced rates of sea level rise and increasing storminess, both of which are associated with global warming. These changes are likely to have significant impact on public health and food security of coastal populations.

A thorough understanding of coastal geological processes is what constitutes the bedrock upon which lies the formulation of policies aimed at addressing issues associated with thresholds and complexities of these processes, and their impacts on population health and food systems. The information garnered in this Chapter, it is hoped, would contribute to vulnerability assessment and definition of adaptation strategies to cope with the impacts of climate change in a more integrated manner.

The production of a catalogue of modern agricultural production technologies and techniques based on modelled patterns of micronutrient flow under climate change scenario is essential. It is concluded that attempts need to be made to quantify the effects of these climate accentuated geological processes on nutrient cycling, and to use this knowledge to formulate guidelines that would help farmers in the selected catchments implement the correct soil remediation practices and adopt other pertinent adaptation strategies to ensure better agricultural productivity and hence, ensure food security.

Finally, it could be said that the increasing public awareness coupled with increasing scientific knowledge, have, over the last 30 years of research, had a major influence on coastal zone management; though admittedly, more research needs to be done. It is hoped that a successful implementation of the recommendations from this discourse would go some way towards bridging the massive knowledge gap that remains in African coastal climate research.

References

ACSS (The Africa Center for Strategic Studies) (2022) Rising Sea Levels Besieging Africa's Booming Coastal Cities. https://africacenter.org/spotlight/rising-sea-levels-besieging-africas-booming-coastal-... Accessed 03 May 2025

Adeyeri OE, Folorunsho AH, Adeliyi TE, Ayegbusi KI, Akinsanola AA, Ndehedehe CE, Ahmed N, Babalola TE (2024) Climate change is intensifying rainfall erosivity and soil erosion in West Africa. Sci Total Environ 955:177174. https://doi.org/10.1016/j.scitotenv.2024.177174. Accessed 05 Mar 2025

Almar R, Stieglitz T, Addo KA, Ba K, Ondoa EA et al (2023) Coastal zone changes in West Africa: challenges and opportunities for satellite Earth observations. Surv Geophys 44:249–275. https://doi.org/10.1007/s10712-022-09721. Accessed 02 Jan 2025

Ankrah J, Monteiro A, Madureira H (2023) Shoreline change and coastal erosion in West Africa: a systematic review of research progress and policy recommendation. Geosciences 13(59). https://doi.org/10.3390/geosciences13020059. Accessed 02 Jan 2025

Archer L, Neal J, Bates P, Lord N, Hawke L, Collings T, Quinn N, Sear D (2024) Population exposure to flooding in Small Island Developing States under climate change. Environ Res Lett 19:124020 (IOP Publishing). https://iopscience.iop.org/article/10.1088/1748-9326/ad78eb/pdf. Accessed 29 Mar 2025

Arteaga E, Nalau J, Biesbroek R, Howes M (2023) Unpacking the theory-practice gap in climate adaptation. Climate Risk Manag 45:100567. https://doi.org/10.1016/j.crm.2023.100567. Accessed 29 Mar 2025

Asmall T, Abrams A, Röösli M, Cissé G, Carden K, Dalvie MA (2021) The adverse health effects associated with drought in Africa. Sci Total Environ 793:148500. https://doi.org/10.1016/j.scitotenv.2021.148500. Accessed 02 Jan 2025

Awuor CB, Orindi VA, Adwera AO (2008) Climate change and coastal cities: the case of Mombasa, Kenya. Environ Urban 20:231–242. https://doi.org/10.1177/0956247808089158. Accessed 13 Mar 2025

Babapoorkamani A, Ricci L (2025) Decision-making strategies for climate change adaptation in coastal regions of Africa. Environ Dev 55:101196. https://doi.org/10.1016/j.envdev.2025.101196. Accessed 29 Mar 2025

Beroya-Eitner MA (2016) Ecological vulnerability indicators. Ecol Indic 60:329–334. https://api.semanticscholar.org/CorpusID:82137908. Accessed 03 Jan 2025

Bertrand S (2021) Fact sheet: climate, environmental and health impacts of fossil fuels. Environmental and Energy Study Institute (EESI). https://www.eesi.org/papers/view/fact-sheet-climate-environmental-and-health-impacts-of-fossil-fuels-2021. Accessed 03 May 2025

Botes A, McKenzie M (2013) Durban climate change strategy health theme report: draft for public comment. Environmental Planning and Climate Protection Department/Energy Office, eThekwini. http://www.durban.gov.za/City_Services/energyoffice/Documents/DCCS%20Health%20Theme%20Report.pdf. Accessed 09 June 2021

Brimicombe C, Jackson D, Mungatia A, Sulaiman Z, Monthaler T, Wieser K, Otto IM (2025) The influence of heat exposure on birth and neonatal outcomes in Mombasa, Kenya: a pooled time series analysis. J Climate Change Health. https://api.semanticscholar.org/CorpusID:275715712. Accessed 13 Mar 2025

Brown O, Flegg JA, Weiss DJ, Golding N (2024) A global mathematical model of climatic suitability for *Plasmodium falciparum* malaria. Malaria J 23(306). https://doi.org/10.1186/s12936-024-05122-7. Accessed 07 Aug 2025

C40 Cities (2021) A renewable energy roadmap for African Cities. https://www.cityenergy.org.za/wp-content/uploads/2021/02/resource_512.pdf. Accessed 03 May 2025

C40 Cities (2022) Benefits of urban climate action. C40 Cities: Climate, Air Quality and Health: Durban. Technical Assistance Report, C40 Cities. https://www.c40.org/wp-content/uploads/2022/02/Durban-%E2%80%93-Regulating-Industrial-Emissions.pdf. Accessed 30 Apr 2025

C40 Cities (2023) A clean and inclusive energy future for African C40 Member Cities. https://sustainable.org.za/wp-content/uploads/2023/11/C40-cities-report-FINAL-…. Accessed 03 May 2025

Ceraldi TS, Hodgkinson R, Backé G (2016) The petroleum geology of the West Africa margin: an introduction. Geological Society of London, Special Publication, vol 438. https://doi.org/10.1144/SP438.11. http://sp.lyellcollection.org/content/early/2016/08/12/SP438.11.full.pdf+html. Accessed 14 Oct 2016

Charnley GEC, Kelman I, Green N, Hinsley W, Gaythorpe KAM, Murray KA (2021) Exploring relationships between drought and epidemic cholera in Africa using generalised linear models. BMC Infectious Dis 21(1):1177. https://doi.org/10.1186/s12879-021-06856-4. Accessed 03 May 2025

Charnley GEC, Kelman I, Murray KA (2022) Drought-related cholera outbreaks in Africa and the implications for climate change: a narrative review. Pathogens Glob Health 116(1):3–12. https://doi.org/10.1080/20477724.2021.1981716. Accessed 03 May 2025

Chapungu L, Nhamo G, Chikodzi D, Dube K (2024) Trends and impacts of temperature and fire regimes in South Africa's coastal national parks: implications for tourism. Nat Hazards 120:4485–4506. https://doi.org/10.1007/s11069-023-06384-1. Accessed 03 Jan 2025

Claessens L, Breugel P, Van Notenbaert A, Herrero M, Van De Steeg J (2008) Mapping potential soil erosion in East Africa using the universal soil loss equation and secondary data. In: Sediment dynamics in changing environments. Proceedings of a symposium held in christchurch, New Zealand, IAHS Publication 325. International Association of Hydrological Sciences, Christchurch (New Zealand), pp 398–407. https://hdl.handle.net/10568/1825. Accessed 05 May 2025

Cook BI, Mankin JS, Anchukaitis KJ (2018) Climate change and drought: from past to future. Curr Climate Change Rep 4:164–179 https://doi.org/10.1007/s40641-018-0093-2. Accessed 30 Dec 2024

Colón-González FJ, Sewe MO, Tompkins AM, Sjödin H, Casallas A, Rocklöv J, Caminade C, Lowe R (2021) Projecting the risk of mosquito-borne diseases in a warmer and more populated world: a multi-model, multi-scenario intercomparison modelling study. Lancet Planetary Health 5(7):e404–e414. https://doi.org/10.1016/S2542-5196(21)00132-7. Accessed 03 May 2025

Costello A, Abbas M, Allen A, Ball S, Bell S, Bellamy R, Friel S, Groce N, Johnson A, Kett M, Lee M, Levy C, Maslin M, McCoy D, McGuire B, Montgomery H, Napier D, Pagel C, Patel J, de Oliveira JA et al (2009) Managing the health effects of climate change: Lancet and University College London Institute for Global Health Commission. Lancet (London, England) 373(9676):1693–1733. https://doi.org/10.1016/S0140-6736(09)60935-1. Accessed 01 Jan 2025

Dada OA, Qiao, L, Ding D, Li G, Ma Y, Wang L (2015) Evolutionary trends of the Niger Delta shoreline during the last 100 years: responses to rainfall and river discharge. Marine Geol 367(C):202–211. https://doi.org/10.1016/j.margeo.2015.06.007. Accessed 05 Mar 2025

Dada OA, Almar R, Morand P (2024) Coastal vulnerability assessment of the West African coast to flooding and erosion. Sci Rep 14. https://doi.org/10.1038/s41598-023-48612-5. Accessed 02 Jan 2025

D'Amato G, Chong-Neto HJ, Monge Ortega OP, Vitale C, Ansotegui I, Rosario N et al (2020) The effects of climate change on respiratory allergy and asthma induced by pollen and mold allergens. Allergy 75(9):2219–2228. https://doi.org/10.1111/all.14476. Accessed 05 May 2025 de Perez EC, Berse KB, Depante LAC, Easton-Calabria E, Evidente EPR, Ezike T (2022) Learning from the past in moving to the future: invest in communication and response to weather early warnings to reduce death and damage. Climate Risk Manag 38:100461. https://doi.org/10.1016/j.crm.2022.100461. Assessed 04 Jan 2025

de Villiers NM, Hodgson AN, Harasti D, Claassens L (2022) The impact of a Reno mattress installation on an adjacent seagrass meadow and its macrofauna in a South African estuary. Regn Stud Marine Sci (52):102239. https://doi.org/10.1016/j.rsma.2022.102239. Accessed 07 Aug 2025

DEWME (Department of Environment, Waste Management and Energy) (2021) Mombasa county: climate change policy. Chrome-extension://efaidnbmnnnibpcajpcglclefindmkaj/https://www.mombasa.go.ke/wp-content/uploads/2021/10/Mombasa-Climate-Change-Policy-2021.pdf. Accessed 11 Mar 2025

Ducruet C, Polo Martin B, Sene MA, Lo Prete M, Sun L, Itoh H, Pigné Y (2024) Ports and their influence on local air pollution and public health: a global analysis. Sci Environ 915:170099. https://doi.org/10.1016/j.scitotenv.2024.170099. Accessed 29 Mar 2025

Echendu AJ (2020) The impact of flooding on Nigeria's sustainable development goals (SDGs). Ecosyst Health Sustain 6(1):1791735. https://doi.org/10.1080/20964129.2020.1791735. Accessed 21 Mar 2025

Enaruvbe, G.O. and Atafo, O.P. (2016). Analysis of deforestation pattern in the Niger Delta region of Nigeria. J Land Use Sci 11(1):113–130. https://doi.org/10.1080/1747423X.2014.965279. Accessed 29 Mar 2025

Elias P, Omojola A (2015) Case study: the challenges of climate change for Lagos, Nigeria. Curr Opin Environ Sustain 13:74–78. https://doi.org/10.1016/j.cosust.2015.02.008. Accessed 17 Mar 2025

Estifanos TK, Fisher B, Galford GL, Ricketts TH (2024) Impacts of deforestation on childhood malaria depend on wealth and vector biology. GeoHealth 8(3):e2022GH000764. https://doi.org/10.1029/2022GH000764. Accessed 03 Jan 2025

Eguiluz-Gracia I, Mathioudakis AG, Bartel S, Vijverberg SJH, Fuertes E, Comberiati P, Cai YS, Tomazic PV, Diamant Z, Vestbo J, Galan C, Hoffmann B (2020) The need for clean air: the way air pollution and climate change affect allergic rhinitis and asthma. Allergy 75(9):2170–2184. https://doi.org/10.1111/all.14177(Accessed30.12.2024)

eThekwini Municipality (2011a) Durban a climate for change: transforming Africa's future. prepared for the COP17/CMP7, Durban, South Africa. https://www.durban.gov.za/storage/Documents/Environmental%20Planning%20%20Climate%20Protection%20%20Publications/Climate/Durban%20A%20Climate%20For%20Change.pdf. Accessed 01 Jan 2025

eThekwini Municipality (2011b) Planet in Peril: adapting to climate change in Durban. COP 17: climate change Poster No. 6. The 2011 United Nations Climate Change Conference (COP 17). https://www.durban.gov.za/storage/Documents/Climate/Planet%20in%20Peril%20Adapting%20To%20Climate%20Change%20In%20Durban.pdf. Accessed 02 May 2025

eThekwini Municipality (2019) Durban Climate Action Plan 2019. Towards Climate Resilience and Carbon Neutrality. https://cdn.locomotive.works/sites/5ab410c8a2f42204838f797e/content_entry5c8ab5851647e100801756a3/5e5e3f71469c8b00a735fbac/files/Climate_Action_Plan_web.pdf#:~:text=inter…. Accessed 17 Mar 2025

eThekwini Municipality (2022a) Durban climate change strategy. Report Version 7. chrome-extension://efaidnbmnnnibpcajpcglclefindmkaj/https://www.durban.gov.za/storage/Documents/Climate/DCCS_Strategy.pdf. Accessed 18 Mar 2025

eThekwini Municipality (2022b) About Ethekwini. https://www.durban.gov.za/pages/government/about-ethekwini#:~:text=Primary%20Sector,sixteen%20fish%20ponds%20in%20place. Accessed 18 Mar 2025

Fan F, Gu X, Luo J, Zhang B, Liu H, Yang H, Wang L (2024) Identification of gully erosion activity and its influencing factors: a case study of the Sunshui River Basin. Plos One 19(11):e0309672. https://doi.org/10.1371/journal.pone.0309672(Accessed05.03.2025)

FGN (Federal Government of Nigeria) (2024) Climate change and health: national vulnerability and adaptation assessment report (2024). chrome-extension://efaidnbmnnnibpcajpcglclefindmkaj/https://www.atachcommunity.com/fileadmin/uploads/atach/Documents/Country_documents/Nigeria_Climate_and_Health_VA_assessment_report_-_FINAL.pdf. Accessed 18 Mar 2025

Firoozi AK, Firoozi AA (2024) Water erosion processes: mechanisms, impact, and management strategies. Results Eng 24:103237. https://doi.org/10.1016/j.rineng.2024.103237. Accessed 07 Aug 2025

Gebrehiwot K (2022) Soil management for food security. In: Jhariya MK et al (eds) Natural resources conservation and advances for sustainability, Chapter 3, pp 61–71. https://doi.org/10.1016/B978-0-12-822976-7.00029-6. Accessed 30 Dec 2024

Gerdes ME, Cruz-Cano R, Solaiman S, Ammons S, Allard SM, Sapkota AR, Micallef SA, Goldstein RER (2022) Impact of irrigation water type and sampling frequency on microbial water quality profiles required for compliance with U.S. In: Food safety modernization act produce safety rule standards. Environmental research, vol 205, p 112480. https://doi.org/10.1016/j.envres.2021.112480. Accessed 21 Aug 2025

Githeko AK, Lindsay SW, Confalonieri UE, Patz JA (2000) Climate change and vector-borne diseases: a regional analysis. Bull World Health Organ 78(9):1136–1147. https://www.scielosp.org/pdf/bwho/v78n9/v78n9a09.pdf. Accessed 21 Aug 2025

Gitton M (2024) 2024 Report on Soil Health in Europe: an urgent call to action against soil degradation. https://www.linkedin.com/pulse/2024-report-soil-health-europe-urgent-call-action-against-gitton-arp6e. Accessed 30 Dec 2024

Goshua A, Sampath V, Efobi JA, Nadeau K (2023) The role of climate change in Asthma. Adv Exp Med Biol 1426:25–41. https://doi.org/10.1007/978-3-031-32259-4_2. Accessed 30 Dec 2024

Grimett L (2024) Climate change mitigation and adaptation: is the South African transport sector ready for change?. Am J Ind Bus Manag 14(4):425–439. https://www.scirp.org/journal/paperinformation?paperid=132594. Accessed 25 Mar 2025

Hategekimana Y, Yu L, Nie Y, Zhu J, Liu F, Guo F (2018) Integration of multi-parametric fuzzy analytic hierarchy process and GIS along the UNESCO World Heritage: a flood hazard index, Mombasa County, Kenya. Nat Hazards: J Int Soc Prevent Mitig Nat Hazards (Springer) 92(2):1137–1153. https://doi.org/10.1007/s11069-018-3244-9. Accessed 30 04 2025

IFRC (International Federation of the Red Cross)/Climate Centre (2021) Climate change impacts on health: Kenya assessment. chrome-extension://efaidnbmnnnibpcajpcglclefindmkaj/https://www.climatecentre.org/wp-content/uploads/RCRC_IFRC-Country-assessments-KENYA.pdf. Accessed 11 Mar 2025

Igwaran A, Kayode AJ, Moloantoa KM, Khetsha ZP, Unuofin JO (2024) Cyanobacteria harmful algae blooms: causes, impacts, and risk management. Water Air Soil Pollut 235:71. https://doi.org/10.1007/s11270-023-06782-y. Accessed 07 Aug 2025

Jung Y-J, Khant NA, Kim H, Namkoong S (2023) Impact of climate change on waterborne diseases: directions towards sustainability. Water 15:1298. https://doi.org/10.3390/w15071298. Accessed 30 Dec 2024

Kafi KM, Barau AS, Aliyu A (2021) The effects of windstorm in African medium-sized cities: an analysis of the degree of damage using KDE hotspots and EF-scale matrix. Intl J Dis Risk Reduct 55:102070. https://doi.org/10.1016/j.ijdrr.2021.102070. Accessed 30 Dec 2024

Kelly AM, Radler RDNN (2024) Does energy consumption matter for climate change in Africa? New insights from panel data analysis. Innov Green Dev 3(3):1–10. https://doi.org/10.1016/j.igd.2024.100132. Accessed 06 Feb 2025

Khelifa R, Mahdjoub H, Samways MJ (2022) Combined climatic and anthropogenic stress threaten resilience of important wetland sites in an arid region. Sci Total Environ 806(Pt 4):150806. https://doi.org/10.1016/j.scitotenv.2021.150806. Accessed 03 Jan 2025

Khojasteh D, Lewis M, Tavakoli S, Farzadkhoo M, Felder S, Iglesias G, Glamore W (2022) Sea level rise will change estuarine tidal energy: a review. Renew Sustain Energy Rev 156. ISSN:1879-0690, 1364-0321. https://doi.org/10.1016/j.rser.2021.111855. Assessed 04 Jan 2025

Kimutai J, Barnes C, Zachariah M, Philip SY, Kew SF, Pinto I, Wolski P, Koren G, Vecchi G, Yang W, Li S, Vahlberg M, Singh R, Heinrich D, Arrighi J, Marghidan CP, Thalheimer L, Kane C, Raju E, Otto FEL (2025) Human-induced climate change increased 2021–2022 drought severity in horn of Africa. Weather Climate Extremes 47:100745. https://doi.org/10.1016/j.wace.2025. 100745.(Accessed05.05.2025)

KNBS (Kenya National Bureau of Standards) (2019) Kenya population and housing census volume I: population by county and sub-county. chrome-extension:// efaidnbmnnnibpcajpcglclefindmkaj/https://housingfinanceafrica.org/app/uploads/VOLUME-I-KPHC-2019.pdf. Accessed 13 Mar 2025

Komolafe AA, Adegboyega SA, Anifowose AY, Akinluyi FO, Awoniran DR (2014) Air pollution and climate change in Lagos, Nigeria: needs for proactive approaches to risk management and adaptation. Am J Environ Sci 10(4):412–423. https://doi.org/10.3844/ajessp.2014.412.423. Accessed 16 Mar 2025

Konko Y, Umaru ET, Nimon P, Adjoussi P, Okhimamhe A, Kokou K (2024) Climate change and coastal erosion hotspots in West Africa: the case of Togo. Regn Stud Marine Sci 77:103691. https://doi.org/10.1016/j.rsma.2024.103691. Accessed 16 Mar 2025

Kruger AC, Pillay DL, van Staden M (2016) Indicative hazard profile for strong winds in South Africa. South African J Sci 112(1–2):01–11. https://doi.org/10.17159/sajs.2016/20150094. Accessed 05 Mar 2025

Laura M, Tartari G, Salerno F, Valsecchi L, Bravi C, Lorenzi E, Genoni P, Guzzella L (2017) Climate change impacts on sediment quality of subalpine reservoirs: implications on management. Water 9:680. https://doi.org/10.3390/w9090680. Accessed 05 May 2025

Levy K, Smith SM, Carlton EJ (2018) Climate change impacts on waterborne diseases: moving toward designing interventions. Curr Environ Health Rep 5(2):272–282. https://doi.org/10.1007/ s40572-018-0199-7. Accessed 21 Mar 2025

Liu C-H, Hung C (2023) Reutilization of solid wastes to improve the hydromechanical and mechanical behaviors of soils—A state-of-the-art review. Sustain Environ Res 33(2468–2039):1–21. https://doi.org/10.1186/s42834-023-00179-6. Accessed 06 Feb 2025

Liu B, Han B, Liang X, Liu Y (2024) Hydrogen production from municipal solid waste: potential prediction and environmental impact analysis. Int J Hydrog Energy 52:1445–1456. https://doi. org/10.1016/j.ijhydene.2023.11.027. Accessed 06 Feb 2025

LSG (Lagos State Government) (2025) Lagos economic outlook: charting a resilient path towards a sustainable future (Unpublished); Lagos unveils 2025 economic devt update, charts path to sustainable growth—Nigerian NewsDirect. Accessed 14 Mar 2025

Mafwele BJ, Lee JW (2022) Relationships between transmission of malaria in Africa and climate factors. Sci Rep 12:14392. https://doi.org/10.1038/s41598-022-18782-9. Accessed 14 Mar 2025

Marijnissen RJC, Kok M, Kroeze C, van Loon-Steensma JM (2021) Flood risk reduction by parallel flood defences—Case-study of a coastal multifunctional flood protection zone. Coastal Eng 167:1–18. Article 103903. https://doi.org/10.1016/j.coastaleng.2021.103903. Accessed 24 Mar 2025

Martel P, Sutherland C (2019) Governing river rehabilitation for climate adaptation and water security in Durban, South Africa. In: Cobbinah PB, Addaney M (eds) The geography of climate change adaptation in urban Africa. Springer International Publishing, pp 355–387. https://doi. org/10.1007/978-3-030-04873-0_13

Martel P, Sutherland C, Hannan S (2022) Governing river rehabilitation projects for transformative capacity development. Water Policy 24(5):778–796. https://doi.org/10.2166/wp.2021.071. Accessed 16 Mar 2025

Martello MV, Whittle AJ (2023) Climate-resilient transportation infrastructure in coastal cities. In: Adapting the built environment for climate change; design principles for climate emergencies. Woodhead Publishing Series in Civil and Structural Engineering, pp 73–108

Mash R (2019) The anthropocene—The biggest threat to health on the African continent. Afr J Primary Health Care Family Med 11(1):1–2. https://doi.org/10.4102/phcfm.v11i1.2151. Accessed 03 Jan 2025

Megersa DM, Luo X-S (2025) Effects of climate change on malaria risk to human health: a review. Atmosphere 16:71. https://doi.org/10.3390/atmos16010071. Accessed 14 Mar 2025

Mehra D, Rael T, Bloem MW (2024) A review of the intersection between climate change, agriculture, health, and nutrition in Africa: costs and programmatic options. Front Sustain Food Syst 8:1389730. https://doi.org/10.3389/fsufs.2024.1389730

MEWR (Ministry of Environment and Water Resources) (2024) Lagos state climate adaptation and resilience plan (LCARP). chrome-extension://efaidnbmnnnibpcajpcglclefindmkaj/https://fsd africa.org/wp-content/uploads/2025/01/Appendix-IV-LCARP-Phase-II-Report_compressed. pdf. Accessed 14 Mar 2025

Mitchell SA (2013) The status of wetlands, threats and the predicted effect of global climate change: the situation in Sub-Saharan Africa. Aquatic Sci 75:95–112. https://doi.org/10.1007/s00027-012-0259-2. Accessed 03 Jan 2025

Mkhize M, Sibanda M (2022) Food insecurity in the informal settlements of Inanda households living with children under 60 months in Ethekwini municipality. Children (Basel, Switzerland) 9(10):1521. https://doi.org/10.3390/children9101521. Accessed 18 Mar 2025

Moore S, Colwell R (2025) Climate change and the resurgence of waterborne diseases: focus on Sub-Saharan Africa. Field Act Sci Rep (27):66–70. http://journals.openedition.org/factsreports/7734. Accessed 24 Mar 2025

Moyo E, Nhari LG, Moyo P, Murewanhema G, Dzinamarira T (2023) Health effects of climate change in Africa: a call for an improved implementation of prevention measures. Eco-Environ Health 2(2):74–78. https://doi.org/10.1016/j.eehl.2023.04.004. Accessed 02 Jan 2025

Mueller W, Zamrsky D, Essink GO, Fleming LE, Deshpande A, Makris KC, Wheeler BW, Newton JN, Narayan KMV, Naser AM, Gribble MO (2024) Saltwater intrusion and human health risks for coastal populations under 2050 climate scenarios. Sci Rep 14(1):15881. https://doi.org/10.1038/s41598-024-66956-4. Accessed 30 Dec 2024

Murumkar AR, Durand MT, Fernández A, Moritz M, Mark BG, Phang SC, Laborde S, Scholte P, Shastry A, Hamilton IM (2020) Trends and spatial patterns of 20th century temperature, rainfall and PET in the semi-arid Logone River basin, Sub-Saharan Africa. J Arid Environ 178:104168. https://doi.org/10.1016/j.jaridenv.2020.104168. Accessed 03 Jan 2025

Naidoo G (2023) The mangroves of Africa: a review. Marine Pollut Bull 190:114859. https://doi.org/10.1016/j.marpolbul.2023.114859. Accessed 07 Aug 2025

Nayak SK, Nandimandalam JR (2023) Impacts of climate change and coastal salinization on the environmental risk of heavy metal contamination along the Odisha coast, India. Environ Res 238:117175. https://doi.org/10.1016/j.envres.2023.117175. Accessed 30 Dec 2024

Ndimele PE, Ojewole AE, Mekuleyi GO, Badmos LA, Agosu CM, Olatunbosun ES, Lawal OO, Shittu JA, Joseph OO, Ositimehin KM, Ndimele FC, Ojewole CO, Abdulganiy IO, Ayodele OT (2024) Vulnerability, resilience and adaptation of Lagos coastal communities to flooding. Earth Sci Syst Soc 4. https://doi.org/10.3389/esss.2024.10087. Accessed 14 Mar 2025

NESG (Nigerian Economic Summit Group) (2023) Climate change impacts on food security and nutritional outcomes in Nigeria: challenges and policy options. The Nigerian Economic Summit Group | Research Document: Climate Change Impacts on Food Security and Nutritional Outcomes in Nigeria: Challenges and Policy Options. Accessed 14 Mar 2025

Nhantumbo B, Dada OA, Ghomsi FEK (2023) Sea level rise and climate change—Impacts on coastal systems and cities. https://doi.org/10.5772/intechopen.113083

Njeru MN, Mwangi E, Gatari MJ, Kaniu MI, Kanyeria J, Raheja G, Westervelt DM (2024) First results from a calibrated network of low-cost $PM_{2.5}$ monitors in Mombasa, Kenya show exceedance of healthy guidelines. GeoHealth 8:e2024GH001049. https://doi.org/10.1029/2024GH001049. Accessed 16 Mar 2025

Nwilo PC, Ibe AC, Adegoke JO, Obiefuna JN, Alademomi AS, Okolie CJ, Owoeye OO et al (2020) Long-term determination of shoreline changes along the coast of Lagos. J Geomat 14(1):72–83. https://www.researchgate.net/publication/342692626_Long-term_determination_of_shorel ine_changes_along_the_coast_of_Lagos#fullTextFileContent. Accessed 30 Apr 2025

Obame-Nkoghe J, Agossou AE, Mboowa G, Kamgang B, Caminade C, Duke DC, Githeco AK, Ogega OM et al (2024) Climate-influenced vector-borne diseases in Africa: a call to empower the next generation of African researchers for sustainable solutions. Infect Dis Poverty 13:26. https://doi.org/10.1186/s40249-024-01193-5. Accessed 30 Dec 2024

Obe OB, Morakinyo TE, Mills G (2023) Assessing heat risk in a Sub-Saharan African Humid city, Lagos, Nigeria, using numerical modelling and open-source geospatial socio-demographic datasets. City Environ Interact 20:100128. https://doi.org/10.1016/j.cacint.2023.100128. Accessed 18 Mar 2025

OCM (National Oceanic and Atmospheric Administration, Office for Coastal Management) (2024) Climate change prediction. https://coast.noaa.gov/states/fast-facts/climate-change.html#:~:text=According%20to%20the%20NOAA%20National,will%20rank%20warmest%20on%20record. Accessed 31 Dec 2024

Odada E, Oyebande L, Oguntola AJ (2005) Lake Chad: experience and lessons learned brief. https://api.semanticscholar.org/CorpusID:131638909. Accessed 03 Jan 2025

O'donoghue, S, Morgan D, Hayley, Haydvogl K (2022) The Durban climate change strategy: lessons learnt from the 2021 strategy review and implementation plan. Town Reg Plan 81:84–96. ISSN 2415-0495. https://doi.org/10.18820/2415-0495/trp81i1.7. Accessed 16 Mar 2025

Office of National Statistics (U.K.) (2023) Investment in Flood Defences, U.K. Census 2021. https://www.ons.gov.uk/economy/economicoutputandproductivity/output/articles/investmentinflooddefencesuk/may2023#main-points. Accessed 24 Mar 2025

Okaka FO, Odhiambo BDO (2019) Health vulnerability to flood-induced risks of households in flood-prone informal settlements in the coastal City of Mombasa, Kenya. Nat Hazards 99:1007–1029. https://doi.org/10.1007/s11069-019-03792-0. Accessed 11 Mar 2025

Okeke GN (2022) The Nigerian perspective of global climate change: a case study of coastal areas of Lagos. Open J Environ Res (OJER) 3(2):38–53. https://doi.org/10.52417/ojer.v3i2.428

Oni T, Lawanson T, Mogo E (2021) The case for community-based approaches to integrated governance of climate change and health: Perspectives from Lagos, Nigeria. J Br Acad 9(s7):7–32. https://doi.org/10.5871/jba/009s7.007. Accessed 14 Mar 2025

Page MJ, Moher D, Bossuyt PM, Boutron I, Hoffmann TC, Mulrow CD, Shamseer L, Tetzlaff JM, Akl EA, Brennan SE, Chou R, Glanville J, Grimshaw JM et al (2021) PRISMA 2020 explanation and elaboration: updated guidance and exemplars for reporting systematic reviews. BMJ (Clin Res Ed.) 372(160). https://doi.org/10.1136/bmj.n160. Accessed 02 Jan 2025

Pai SJ, Carter TS, Heald CL, Kroll JH (2022) Updated World Health Organization air quality guidelines highlight the importance of non-anthropogenic $PM_{2.5}$. Environ Sci Technol Lett 9(6):501–506. https://doi.org/10.1021/acs.estlett.2c00203. Accessed 30 Apr 2025

Parums DV (2024) A review of the increasing global impact of climate change on human health and approaches to medical preparedness. Med Sci Monitor: Int Med J Exp Clin Res 30:e945763. https://doi.org/10.12659/MSM.945763. Accessed 30 Dec 2024

Reiner RC, Graetz N Jr, Casey DC, Troeger C, Garcia GM, Mosser JF, Deshpande A, Swartz SJ, Ray SE, Blacker BF, Rao PC et al (2018) Variation in childhood diarrheal morbidity and mortality in Africa, 2000–2015. New Engl J Med 379(12):1128–1138. https://doi.org/10.1056/NEJMoa1716766. Accessed 21 Aug 2025

Roa OH, O'Connor J, Ogunwumi TS, Ihinegbu C, Lynggaard JR, Sebesvari Z, Eberle C, Koli M (2022) Lagos floods. United Nations University, Technical Report TR_220824_LagosFloods. https://digitallibrary.un.org/record/4068191?ln=en&v=pdf. Accessed 21 Aug 2025

Roberts D, Morgan D, O'Donoghue S, Guastella L, Hlongwa N, Price P (2016) Durban, South Africa. In: Bartlett S, Satterthwaite D (eds) Cities on a finite planet: towards transformative responses to climate change, 1st edn., Chapter 6. Routledge, pp 96–115. https://www.researchgate.net/profile/Aromar-Revi/publication/312324494_Bangalore_India_in_Cities_on_a_Finite_Planet.... Accessed 14 Mar 2025

Romanello M, Napoli CD, Green C, Kennard H, Lampard P, Scamman D, Walawender M, Ali Z, Ameli N, Ayeb-Karlsson S et al (2023) The 2023 report of the Lancet Countdown on health and climate change: the imperative for a health-centred response in a world facing irreversible harms. Lancet (London, England) 402(10419):2346–2394. https://doi.org/10.1016/S0140-673 6(23)01859-7. Accessed 24 Mar 2025

Salhi A, El Hasnaoui Y, Pérez Cutillas P, Heggy E (2023) Soil erosion and hydroclimatic hazards in major African port cities: the case study of Tangier. Sci Rep 13:13158. https://doi.org/10.1038/ s41598-023-40135-3. Accessed 05 Mar 2025

Salvador C, Nieto R, Kapwata T, Wright CY, Reason C, Gimeno L, Vicedo-Cabrera AM (2024) Analyzing the effects of drought at different time scales on cause-specific mortality in South Africa. Environ Res Lett 19(5):054022. https://doi.org/10.1088/1748-9326/ad3bd2. Accessed 02 Jan 2025

Sam K, Zabbey N, Gbaa ND, Ezurike JC, Okoro CM (2023) Towards a framework for mangrove restoration and conservation in Nigeria. Regn Stud Marine Sci 66:103154. https://doi.org/10. 1016/j.rsma.2023.103154. Accessed 03 Jan 2025

Sama MA-W, Bérenger V (2023) In Africa, greenhouse gas emissions from the waste sector increase, despite efforts from various actors. Observatory of Climate Action in Africa. https://www.climate-chance.org/wp-content/uploads/2023/06/obsaf_emag6_note-economie-circulaire-et-dechets_eng-2.pdf. Accessed 03 Jan 2025

Same NN, Yakub AO, Chaulagain D, Tangoh AF, Nsafon BEK, Owolabi AB, Suh D, Huh JS (2024) The future of clean energy: agricultural residues as a bioethanol source and its ecological impacts in Africa. Renew Energy 237:121612. https://doi.org/10.1016/j.renene.2024.121612. Accessed 03 Jan 2025

Séranne M, Anka Z (2005) South Atlantic continental margins of Africa: a comparison of the tectonic versus climate interplay on the evolution of equatorial west Africa and SW Africa margins. J Afr Earth Sci 43:283–300. https://doi.org/10.1016/j.jafrearsci.2005.07.010. Accessed 02 Jan 2025

Shaffril HAM, Samah AA, Samsuddin SM, Ahmad N, Tangang F, Sidique SFA, Rahman HA, Burhan NAS et al (2024) Diversification of agriculture practices as a response to climate change impacts among farmers in low-income countries: a systematic literature review. Clim Serv 35:100508. https://doi.org/10.1016/j.cliser.2024.100508. Accessed 05 Mar 2025

Singh P, Saran S (2024) Identifying and predicting climate change impact on vector-borne disease using machine learning: case study of *Plasmodium falciparum* from Africa. In: The international archives of the photogrammetry, remote sensing and spatial information sciences, Volume XLVIII-2-2024 ISPRS TC II Mid-term Symposium "The Role of Photogrammetry for a Sustainable World", Las Vegas, Nevada, USA. Isprs-archives-XLVIII-2-2024-387-2024.pdf. Accessed 07 Aug 2025

SLoCaT (Partnership on Sustainable Low Carbon Transport) (2018) Transport and climate change 2018: global status report. https://www.changing-transport.org/wp-content/uploads/slocat_tra nsport-and-climate-change-2018-web.pdf. Accessed 29 Mar 2025

Stuart J, Yozell S, Ochanda V, Rouleau T, Indasi V, Lombardo K (2021) CORVI risk assessment: Mombasa, Kenya: a holistic city-based assessment of the climate risks facing Mombasa, Kenya. Resilience and Sustainability https://www.stimson.org/2021/corvi-risk-profile-mom basa-kenya/. Accessed 11 Mar 2025

Suhr F, Steinert JI (2022) Epidemiology of floods in Sub-Saharan Africa: a systematic review of health outcomes. BMC Public Health 22(1):268. https://doi.org/10.1186/s12889-022-12584-4. Accessed 03 May 2025

Sutherland C (2024) The direct and indirect climate change impacts, loss and damages in selected at-risk communities in Durban as a result of the April 2022 floods. Centre for Environmental Rights. CER-Report-Sutherland-FINAL-28-Feb-2024.pdf. Accessed 16 Mar 2025

Tajudeen TT, Omotayo A, Ogundele FO, Rathbun LC (2022) The effect of climate change on food crop production in Lagos State. Foods (Basel, Switzerland) 11(24):3987. https://doi.org/ 10.3390/foods11243987. Accessed 14 Mar 2025

Tang H, Du L, Xia C, Luo J (2024) Bridging gaps and seeding futures: A synthesis of soil salinization and the role of plant-soil interactions under climate change. iScience 27(9):110804. https://doi. org/10.1016/j.isci.2024.110804. Accessed 30 Dec 2024

Teku D, Workie MD (2025) Longitudinal analysis of soil erosion dynamics using the RUSLE model in Ethiopia's Lake Ziway watershed: implications for agricultural sustainability and food security. Front Media S.A., Section on Land Use Dynamics 12. https://www.mdpi.com/2220-9964/14/1/28. Accessed 07 Aug 2025

Tully KL, Gedan KB, Epanchin-Niell R, Strong AL, Bernhardt ES, BenDor TK, Mitchell M, Kominoski JS, Jordan TE, Neubauer SC, Weston N (2019) The invisible flood: the chemistry, ecology, and social implications of coastal saltwater intrusion. BioScience 69(5):368–378. https://doi.org/10.1093/biosci/biz027. Accessed 29 Mar 2025

UNU (United Nations University) (2023) Frequent flooding in African coastal cities demand holistic recovery pathways. UNU, Institute for Environment and Human Security. https://unu.edu/ehs/news/frequent-flooding-african-coastal-cities-demand-holistic-recovery-pathways. Accessed 05 May 2025

Varela C, Young S, Mkandawire N, Groen RS, Banza L, Viste A (2019) Transportation barriers to access health care for surgical conditions in Malawi: a cross sectional nationwide household survey. BMC Public Health 19(1):264. https://doi.org/10.1186/s12889-019-6577-8. Accessed 29 Mar 2025

Vousdoukas MI, Athanasiou P, Giardino A, Mentaschi L, Stocchino A, Kopp RE et al (2023) Small Island Developing States under threat by rising seas even in a 1.5 °C warming world. Nat Sustain 6(12):1552–1564. https://doi.org/10.1038/s41893-023-01230-5. Accessed 29 Mar 2025

Vrieling A, Hoedjes JC, Velde MV (2014) Towards large-scale monitoring of soil erosion in Africa: accounting for the dynamics of rainfall erosivity. Glob. Planetary Change 115:33–43. https://doi.org/10.1016/j.gloplacha.2014.01.009. Accessed 05 Mar 2025

Wang L, Cui S, Li Y, Huang H, Manandhar B, Nitivattananon V, Fang X, Huang W (2022) A review of the flood management: from flood control to flood resilience. Heliyon 8(11):e11763. https://doi.org/10.1016/j.heliyon.2022.e11763. Accessed 24 Mar 2025

Wang J, Cortes-Ramirez J, Gan T, Davies JM, Hu W (2024) Effects of climate and environmental factors on childhood and adolescent asthma: a systematic review based on spatial and temporal analysis evidence. Sci Total Environ 951:175863. https://doi.org/10.1016/j.scitotenv.2024.175863. Accessed 30 Dec 2024

Wannewitz M, Ajibade I, Mach KJ, Magnan A, Petzold J, Reckien D, Ulibarri N, Agopian A, Chalastani VI, Hawxwell T et al (2024) Progress and gaps in climate change adaptation in coastal cities across the globe. Nat Cities 1:610–619. https://doi.org/10.1038/s44284-024-00106-9. Accessed 03 Jan 2025

Watts N, Adger WN, Agnolucci P, Blackstock J, Byass P, Cai W, Chaytor S, Colbourn T, Collins M, Cooper A, Cox PM, Depledge J, Drummond P, Ekins P, Galaz V, Grace D, Graham H et al (2015) Health and climate change: policy responses to protect public health. Lancet (London, England) 386(10006):1861–1914. https://doi.org/10.1016/S0140-6736(15)60854-6(Accessed04.01.2025)

WHO (World Health Organisation) (2022) Africa faces rising climate-linked health emergencies. WHO, Geneva. https://www.afro.who.int/news/africa-faces-rising-climate-linked-health-emergencies. Accessed 31 Dec 2024

WHO (World Health Organisation) (2023) Radiation and Health. Geneva. Radiation and health. Accessed 06 Feb 2025

WMO (World Meteorological Organisation) (2024) State of the Climate in Africa 2023. WMO-No. 1360. https://www.developmentaid.org/api/frontend/cms/file/2024/12/1360_State-of-the-Climate-in-Africa-2023_en.pdf. Accessed 03 Jan 2025

World Bank Group (2022) What you need to know about food security and climate change. https://www.worldbank.org/en/news/feature/2022/10/17/what-you-need-to-know-about-food-security-and-climate-change#:~:text=Use%20water%20more%20efficiently%20and,This%20is%20hugely%20important. Accessed 18 Mar 2025

Wright CY, Kapwata T, Naidoo N, Asante KP, Arku RE, Cissé G, Simane B, Atuyambe L, Berhane K (2024) Climate change and human health in Africa in relation to opportunities to strengthen mitigating potential and adaptive capacity: Strategies to inform an African "Brains Trust". Annal Glob Health 90(1):7. https://doi.org/10.5334/aogh.4260. Accessed 05 Mar 2025

Yussuf E, Muthama JN, Mutai B, Marangu DM (2023) Impacts of air pollution on paediatric respiratory infections under a changing climate in Kenyan urban cities. East Afr J Sci Technol Innov 4(2). https://doi.org/10.3329/bjar.v42i3.34505. Accessed 16 Mar 2025

Zenda M (2024) A systematic literature review on the impact of climate change on the livelihoods of smallholder farmers in South Africa. Heliyon 10(18):e38162. https://doi.org/10.1016/j.heliyon.2024.e38162. Accessed 05 Mar 2025

Zhang X, Chen P, Han Y, Chang X, Dai S (2022) Quantitative analysis of self-purification capacity of non-point source pollutants in watersheds based on SWAT model. Ecol Indicat 143:109425. https://doi.org/10.1016/j.ecolind.2022.109425. Accessed 07 Mar 2025

Zhang Z, Chen Z, Zhang J, Liu Y, Chen L, Yang M, Osman AI, Farghali M, Liu E, Hassan D, Ihara I, Lu K, Rooney DW, Yap PS (2024) Municipal solid waste management challenges in developing regions: a comprehensive review and future perspectives for Asia and Africa. Sci Total Environ 930:172794. https://doi.org/10.1016/j.scitotenv.2024.172794. Accessed 06 Feb 2025

Zhou L, Shen G, Li C, Chen T, Li S, Brown R (2021) Impacts of land covers on stormwater runoff and urban development: a land use and parcel-based regression approach. Land Use Policy 103:105280. https://doi.org/10.1016/j.landusepol.2021.105280. Accessed 07 Aug 2025

Zi Z, Ji D, Jie L, Di W, Guanghao (2023) Enhancing energy–climate–economy sustainability in coastal cities through integration of seawater and solar energy. Coastal Cities. https://doi.org/10.1016/j.rser.2023.113477. Accessed 02 Jan 2025

Chapter 7
A Code for Medical Geology Fieldwork in Africa: Guidelines on Health and Safety Issues in Mapping Disease Distribution and Their Geoenvironmental Correlates

D. S. Nadasan and T. C. Davies

Abstract Sometimes, a rare and strictly locality related disease outbreak occurs in some African countries, the aetiology of which cannot be readily and clearly established. In such cases, a suspected geoenvironmental factor or cofactors is/are often implicated. In handling such situations, Medical Geologists are often summoned to work in teams that include epidemiologists, public health specialists and toxicologists, who investigate the problem so as to identify causes and risk factors, and implement prevention and control measures. Drawing correlations between disease distribution and some geoenvironmental cofactor(s) could involve substantial amount of fieldwork, whereby Medical Geologists, just like the other specialists in the team, may be exposed to various hazards. Examples of such exposures include, during data collection on patterns of silica dust emission and distribution; sampling of *geophagic* materials for microbiological analyses, and sampling of radioactive tailings for determining uranium migration pathways and particle concentrations in air. The first step in such investigations is to construct a health and safety plan (HSP), whose purpose is to provide a means for minimising accidents and injuries that may occur at a specific location or while working on a specific project; and to communicate to all involved what safety procedures are to be followed.

Keywords Medical geology · Field surveys · Health · Safety · Guidelines

D. S. Nadasan · T. C. Davies (✉)
Faculty of Applied and Health Sciences, Mangosuthu University of Technology, Umlazi,
KwaZulu Natal Province, Republic of South Africa
e-mail: daviestheophilus2025@yahoo.com

© The Author(s), under exclusive license to Springer Nature Switzerland AG 2026
T. C. Davies (ed.), *Recent Advances in Medical Geology Research
in Africa: A Decadal View*, SpringerBriefs in Earth System Sciences,
https://doi.org/10.1007/978-3-032-13754-8_7

Introduction

Medical Geology is still a relatively new and evolving field of study, research and fieldwork. A generic definition of fieldwork undertaken by university staff or student is: "any off-site work carried out for the purpose of teaching, research or other activities under the aegis of the institution" (University of Bath 2025). Although Medical Geology fieldwork fits into this definition, it does bear some uniqueness, for it synthesises field observations using techniques from several allied disciplines in order to reach tangible conclusions. Fieldwork in Medical Geology takes place in a range of settings, engendering methodological considerations of risk and safety (Davies 2018). This Chapter offers a sketch of health and safety issues that the Medical Geologist can encounter during field expeditions, how such work affects researchers and how health and safety issues can be mitigated or obviated. The Chapter incorporates a rich and diverse set of examples depicting fieldwork experiences, insights and reflections on conducting Medical Geology fieldwork, the health and safety issues that are likely to be encountered, the solutions that could be developed, and the realities of "being in the field".

Preparation for Fieldwork

Among the primary objectives of this Chapter is the spelling out of the logistical planning and execution that is required to undertake a successful Medical Geology field project; in particular, from the point of view of health and safety. A well-drawn research plan would help the Medical Geology researcher avoid disaster and equip him/her to deal with it when it happens.

Incorporated in the fieldwork schedule, is a field sampling plan (FSP). A well-drawn FSP should provide comprehensive guidance for sample collection, preparation, preservation and shipping arrangements. Tracking of field samples and recording field data should be incorporated in the plan. The FSP should spell out the Analytical Protocol Specifications (APSs) and the measurement quality objectives (MQOs) that must be met.

When discussing your research with your research group, please think about the health and safety implications of where you are going and what you are doing. After this discussion and before your travel is approved by your institution, you will be required to complete a "Fieldwork Risk Assessment Form" which is normally issued by your Health and Safety Office Administrator or similar officer. Sound plans for averting disaster on field trips should include staying in top physical condition (See: Davies 2018). Fitness and advance planning can help one navigate the subsequent gruelling challenges that sometimes await the Medical Geologist in the African wilds.

Working with Interdisciplinary Colleagues

The world is rapidly evolving, and tackling problems in accurate medical diagnosis is among the many complex challenges that require fresh and comprehensive approaches to solve them (See, e.g., Levain 2023). In no other discipline is the strength of the interdisciplinary approach more exquisitely demonstrated than in Medical Geology. This emerging field offers a powerful multidisciplinary approach that draws upon the collective insights, methodologies, and expertise from diverse fields such as epidemiology, geochemistry, toxicology and biogeochemistry to address diagnostic uncertainties. Effective Medical Geology fieldwork also draws from methodologies and field techniques from those employed by the various specialists that collaborate in reaching tangible conclusions for improvement of medical diagnoses.

The multidisciplinary approach also encourages collaboration and fosters innovation, enabling the researchers to address problems from multiple perspectives and unlock transformative solutions. Experts have for long been working within the confines of their own silos, often incognisant of new developments and insights made in other disciplines. Today, it is well established that leveraging the collective expertise of various disciplines does allow for a more comprehensive understanding of complex problems and offers the opportunity to uncover connections and solutions that may otherwise remain hidden (See: Levain 2023).

The Risk Management Process: Mitigating Risk in Medical Geology Fieldwork

Fieldwork is regarded as being successful, if it achieves the laid out research or scholarly objectives without having to face intractable health and safety challenges. So, it is important to have the *risk management process* in place. *Risk identification* and *risk assessment* are essential stages of the risk management process. Risk assessments are the series of steps taken to identify potential hazards (anything that can cause harm) associated with the work, what risks these hazards may pose, and ways to reduce or eliminate them.

Risks can come from various sources, and its potential impact can be characterised as: "Insignificant", "Minor", "Moderate", "Major" or "Extreme/Catastrophic" based on the level of severity (Fig. 7.1). The University of Stellenbosch (UoS) of South Africa which offers dedicated programmes on "The Risk Management Process" (UoS 2024), recognises three basic categories of risk: "Low", "Medium" or "High" risk, depending on the combinations of likelihood and consequences, as follows:

- Low risk: Research activities in which the only foreseeable risk is one of discomfort or inconvenience.

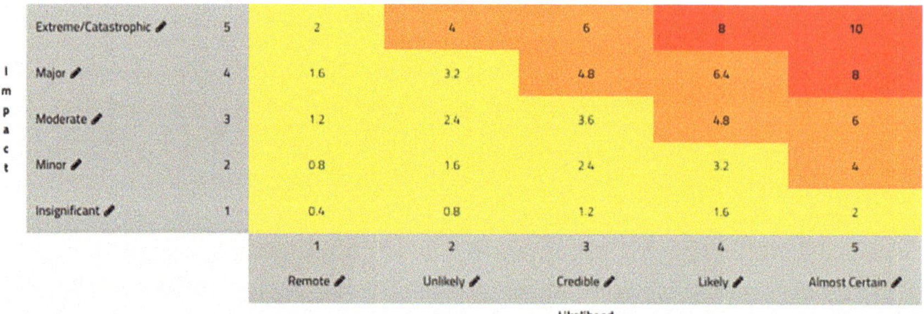

Fig. 7.1 The OWASP (Open Worldwide Application Security Project) risk rating methodology and simplerisk. From: https://www.simplerisk.com/sites/default/files/inline-images/Screen%20S hot%202021-02-25%20at%2010.08.02%20AM.png (Accessed 05.08.2025). Credit: Josh Sokol (Simplerisk) (2024); Permission for reuse granted on 05.08.2025

- Medium risk: Research activities in which an individual or group is exposed to key sources of vulnerability (social, psychological, economic, legal) in the context of the research project.
- High risk: Research activities activities which might place either the participant(s) or the researcher(s) at risk of harm (UoS, Undated).

Then follows the proactive actions taken by researchers to prevent or reduce to an acceptable level, the impact of the potential risk events on the fieldwork.

Health and Safety Precautions in Sampling Dust

Airborne dust is produced from a wide variety of natural (geogenic) processes: mines and quarries (dust containing crystalline silica, coal dust, dust containing metals such as cadmium and lead, and asbestos) (See: Chap. 5, this Volume). There are also several anthropogenic sources, such as from agriculture and forestry activities and housing developments, tunnelling, and glass and ceramics manufacturing. Large quantities of dust can also be generated from other natural sources such as dry riverbeds, earthquakes and volcanic eruptions. The growing awareness that dust presents a significant health and safety hazard has prompted regulatory responses from some African governments, although there are many instances where monitoring and enforcement of regulatory guidelines are ineffective.

Sampling of Geogenic Dust

Dust sampling refers to the gathering of evidence, a systematic collection of data that define the air quality within a professional setting. At the essence of dust sampling is the meticulous collection and analysis of air samples. It is a methodical approach that requires precision instruments and trained personnel to ensure the accuracy of the findings.

The geology of the rock and kind of methods used to extract and process the ores will determine the type and quantity of dust particles generated, and the requisite tools for sampling.

The tools employed in dust sampling are quite varied. *Air sampling pumps* are the engines that drive the sampling process, drawing air through filters that act as nets capturing dust particles for subsequent analysis. *Filter cassettes* then safeguard these captured particles, while *flow meters* serve as the timekeepers, ensuring that air flows through the system at the right pace for accurate measurement.

What Damage Can Dusts Cause?

Dusts generally cause damage to the lungs and respiratory system, but some types can cause cancer (See Chap. 5, this Volume). The main diseases associated with inhalation of hazardous dusts are: *pneumoconiosis* (inhalation of fine particulate silica dust or coal dust into the airways and lung tissue reactions to its presence), including *silicosis, pulmonary fibrosis, coal workers' pneumoconiosis* (CWP) or *black lung disease (from coal dust), asbestosis* (from asbestos dust), *mesothelioma of pleura* and *lung cancer, talcosis* (from talc dust); *siderosis* (from iron dust); and *inflammation of the lung tissues* or *bronchioles* mainly caused by inhalation of certain metal dusts (e.g., cadmium and beryllium dusts), and similar to pneumonia, but vary in severity, depending on which metal is inhaled (See: Chap. 5, this Volume).

Personal Protective Measures When Sampling Dust

When undertaking dust sampling, it is necessary to have access to disposable personal protective equipment (PPE) and clothing, when working in environments that cause or raise up dust, regardless of whether or not dust can be seen. Breathable fabrics and masks provide an excellent particulate barrier. Protective eyewear or goggles are also recommended. Gloves and other skin protection must be worn if dust absorption through the skin or through ingestion, is considered likely; or if dust can have a direct effect on the skin.

Finally, PPE and especially, respiratory protective equipment (RPE) must be thoroughly cleaned and well maintained, for these equipment items to remain effective. Poor maintenance makes any PPE ineffective.

Guidelines on Sampling for Radiological Monitoring

Just as in other types of field sampling programme, but more so in sampling for radiological monitoring (radioactive materials), a FSP is an absolute essential (See section on: "Preparation for Fieldwork", this Chapter). This section provides guidance on health and safety aspects in sampling of radioactively contaminated and mixed contaminated soils and other ground material such as waste rock material and sediments and the types of PPE that are commonly used.

The document widely recognised as providing relevant information on safety issues relating to the sampling of soil and other ground material is the ISO Document on: "Soil Quality Sampling, Part 103" (ISO 2017, Updated 2022), to which the reader is referred. This Part (of the ISO Document) states, *inter alia*:

> The main objectives of this guidance on safety, are (a) to identify the hazards that may exist in carrying out site investigations and soil sampling programmes, (b) to indicate management procedures to provide a framework for safe working and proper response in the case of accident, (c) to indicate what precautions can be taken in terms of personal protection and cleaning facilities to minimize any hazard, and (d) to indicate what working procedures can be adopted to minimize hazards from contaminants and physical hazards associated with the collection of samples and the use of machinery.

Because the ISO (2017) Document was updated in 2022, it represents more recent information than what is given in Barnekow et al. (2019) work on: "Guidelines on Soil and Vegetation Sampling for Radiological Monitoring", which is also worthy of review.

Airborne Radioactivity

Radiological air sampling is used to measure the amount of radioactive materials suspended in the air. The main risk posed in sampling that involves airborne radioactivity is from internal irradiation following inhalation.

Airborne radioactivity measurements undertaken to assess exposures of researchers and members of the general public commonly involve the use of portable or stationary air pumps/samplers, but passive air samplers are sometimes used where appropriate.

Use of PPE in Radiological Monitoring

Personnel safety requirements and considerations are of especial concern in radiological monitoring. They must be worn as stipulated by radiation protection personnel and the site safety officer. Such equipment items (PPE) should be used during all sample collection and decontamination activities (Table 7.1).

Table 7.1 Use of personal protective equipment in radiochemical analyses[a]

Personal protective equipment	Description and use
Boot/shoe covers	These are made from plastic and are meant to protect the wearer's shoes and the environment from contamination, keeping them dry, clean, and free from dirt, mud, dust, and liquids
Coveralls	These are one-piece protective garments made from Paper-Tyvek®. This is a non-woven, synthetic material composed of high-density polyethylene fibres that resemble paper. Tyvek® garments provide an ideal balance of protection, durability and comfort
Cotton coveralls	One-piece protective garments made of cotton, which is known for its breathability, comfort, and absorbency. Cotton coveralls are designed to provide maximum comfort for industrial (re: radiation-) workers
Chemical resistant coveralls	One-piece protective garments providing a critical barrier against penetration of hazardous substances like radioactive materials, liquid splashes, sprays, aerosols, or dry particulates; and offering mechanical resistance against abrasion, punctures or tears
Ear protection	Ear protection devices, such as earplugs, earmuffs, and canal caps, are designed to protect the ears from excessive noise However, radiochemical analysis itself, being a laboratory-based procedure, hardly produces noise pollution. Radiochemical analytical facilities though, are often located in buildings or research laboratories where external noise pollution from sources like construction or heavy machinery could potentially occur
Eyewear	Exposure to low-dose radiation can lead to conditions like cataracts, glaucoma, and retinal damage Radiation eyewear, specifically lead-lined radiation safety glasses, along with other PPE is a key component of the necessary safety measures in a radiochemistry laboratory; and should be highly recommended in order to protect the eyes from potential radiation hazards. Radiation glasses are carefully engineered with features like lead lining, ergonomic frames, and side shields to ensure maximum safety and protection from radiation hazards. These glasses provide crucial shielding against prolonged exposure to ionising radiation, such as X-rays, by absorbing and blocking the radiation from reaching the eyes, which are highly sensitive to damage For professionals working in cath labs (catheterisation laboratories) and outdoor workers in bright conditions, the use of radiation glasses is of utmost importance since these individuals are exposed to low-dose radiation Using radiation glasses helps in following the ALARA (As Low As Reasonably Achievable) principle to minimise radiation exposure
Face shields	Face shields, often made of lead or lead acrylic, protect the eyes and face from harmful ionising radiation, particularly scatter radiation from X-ray procedures and beta particles from various radioactive sources. These shields provide a protective barrier, supplementing other radiation protection measures like lead glasses and distance, and are used in medical, research, and industrial settings where there may be risks of radiation exposure

(continued)

Table 7.1 (continued)

Personal protective equipment	Description and use
Exam gloves	*Exam gloves* made from latex or nitrile (powder free) are a kind of personal protective equipment worn by personnel to shield against radioactive contamination and chemical exposure during radiochemical analyses. Studies have been conducted to evaluate the suitability of various glove materials (e.g., nitrile), for radiopharmaceutical protection, and to assess their permeability to radionuclides like technetium-99 m and fluorine-18. Such evaluations are essential for radiation protection, and for ensuring that the gloves effectively prevent skin contamination and dose accumulation to the hands of operators in nuclear medicine and radiochemistry settings
Radiation hats	Radiation hats are a type of hard hat for radiation protection. These hats are designed to provide exceptional shielding against harmful radiation. They are a critical component in safeguarding individuals who work in radiation-exposed environments. These hats are designed to ensure comprehensive coverage for the head, a critical area often exposed during radiation-related tasks
Respirators	Respirators are used in radiochemical analysis to protect workers from airborne hazardous substances by providing clean air to breathe through filters and cartridges designed to capture specific contaminants. While general respirators protect against dusts, mists, gases, and vapours, radiological hazards require specialised filters, and chemical hazards require corresponding cartridges. Choosing the correct respirator for radiochemical analysis depends on the specific radioactive particles and chemical substances present
Monitoring devices	A radiation dosimeter lapel sampler collects air samples near a worker's breathing zone; and the filter media is then sent for radiochemical analysis to measure and identify airborne radioactive materials. This procedure provides data needed for assessing internal radiation dose. The radiochemical analysis uses techniques such as gamma spectrometry to determine the type and quantity of radionuclides in the sample, helping to evaluate potential internal exposure risks to the worker

[a] Authors' original compilation

The minimal amount of PPE typically used include:

- Coveralls.
- Gloves.
- Waterproof or water-resistant shoe covers or boots. See: Table 7.1 for a comprehensive listing of the full complement of PPE required.

Control and Prevention

Information on controlling ionising radiation hazards and reducing dose is a necessary inclusion in the FSP. A radiation protection programme should normally be managed by a qualified expert such as a health physicist, who is often called a radiation safety officer (RSO). A dosimetry programme is one in which monitoring of personal exposure is conducted, as required by national regulations, *for external dose* (radiation dose received from outside the body) and, as needed, for *internal dose* (radiation dose received from radioactive material deposited inside the body).

A major requirement as stipulated in radiation protection programmes is keeping each worker's occupational radiation dose "As Low As Reasonably Achievable" (ALARA). One of the most essential functions of a radiation protection programme is training radiation workers on safe work practices.

Survey Instruments

Radiation survey instruments can be used to measure exposure rates, dose rates, and the quantities (activity) of radioactive materials and contamination. The choice of survey instrument must be appropriate for the type and energy of the radiation being measured. The most commonly used and recognisable survey meters for measuring ionising radiation are hand-held types. *Radioisotope Identification Devices* (RIIDs) are hand held radiation instruments designed to identify the radioactive isotopes in a radiation source. Personal Radiation Detectors (PRD) are small electronic devices designed to alert the wearer to the presence of radiation.

Continuous air monitors (CAM) can be used to evaluate the presence of airborne radioactive material. These devices can be used to alert personnel to an increased level of radioactive material in the air that may require some action, such as evacuation.

Concentration of radon in air can be carried out using several different methods. The use of *diffusive samplers* to measure the average airborne radon concentration can be made over extended sampling periods. Commercially available radon test kits are now readily available and are an example of a diffusive type of sampler.

Analysis of Radioactive Materials

Radioactive samples can be analysed using a variety of equipment types depending on the type of sample (e.g. air, water, soil, surface wipe) and the types of radiation emitted by the sample.

Examples of the Types of Equipment Used to Evaluate Radioactive Samples

Scaler and *Counters* are devices that are often portable and are used to measure the amount of alpha or beta radiations on a radiological sample.

Liquid Scintillation Counters are not portable and are usually deployed in a laboratory. This instrument can be used for all types of radiations, but it is most often used for measuring beta particles.

Gamma spectroscopy is a method used to identify the radioisotopes present in a radiological sample and quantify the amount of radioactivity in that sample. While these devices can be handheld like the RIID, the most sensitive and accurate instruments are not portable and are only used in the laboratory.

Alpha Spectroscopy is a technique used to identify and quantify alpha emitting radioisotopes. Radioactive samples are chemically digested and the solution is placed onto a thin metal disk for further analysis.

Whole-Body Counting

A *whole-body counter* (WBC) is an extremely sensitive detector, or series of detectors, used to measure the amount of radioactivity inside a human body or animal body. These instruments rely on the measurement of gamma and x-rays emitted from the radioactive material deposited in the body.

Bioassay Sampling

Bioassay sampling (e.g., of urine, faeces, and blood) is conducted to determine the uptake of radioactive material by researchers doing radiological sampling. Samples are typically collected at the beginning of the sampling exercise, periodically during the exercise, after known or suspected intakes, and at the termination of sampling programme, to determine occupational radiation doses.

Sampling in Remote, Hazardous and Extreme Environments

Remote fieldwork refers to all rural fieldwork but is further characterised in terms of distance and accessibility. As of 2024, over fifty percent of Africa's population was rural (Statista 2024), depending on water and food sources from their immediate vicinity for their daily sustenance (Fig. 7.2). Living so close to the land, in the context of the source of their livelihoods, coupled with the geochemical uniqueness and complexity of Africa's surface geochemical environment, Medical Geology research holds much more promise in the Continent's rural and remote settings in the context of obtaining the most meaningful and interpretable results; hence, making them (these areas) attractive to Medical Geology researchers.

However, sampling in remote locations exposes researchers to a range of hazards associated with the type of activity undertaken. The same can be said about working in hazardous and extreme environments such as landmine fields, nuclear sites, waste disposal sites, mountainous settings and slopes and quarries, and the Medical Geologist has to be informed and made aware of these hazards in order to be able to plan and take steps to mitigate the risk.

A good knowledge and awareness of the potential hazards can help you plan and take steps to mitigate the risk. The importance of assessing the terrain and the potential risks of completing the work successfully in such extreme environments should be incorporated in terrain risk assessments within project planning (See section on: "The Risk Management Process", this Chapter). Being prepared is the key to survival!

The sheer number and variety of concerns that need to be addressed makes it unwise for researchers to work alone in such situations. For both safety and scientific discussions onsite, it is recommended that fieldwork be conducted by groups of at least two people; and the party to be led by someone familiar with the hazards to be encountered and the precautions to be taken to minimise risks. An employee of the Ministry of Defence (MoD), Ministry of the Environment (MoE), The National

Fig. 7.2 Over fifty percent of Africa's population lives in a rural setting, depending on water and food sources from their immediate vicinity for their daily sustenance. From: Healy, T. (2018). Why have food security projects failed in rural Africa? https://www.thebigq.org/2018/04/09/why-are-afr icans-still-, (Accessed 24.08.2025). Credit: Terrence Healy, University of Newcastle. Reproduced with permission granted on 27.08.2025

Disaster Response Department (NDRD), mine or quarry, or an expert guide on the natural features, or the fieldwork leader, are among those suitable for performing this role.

In mountainous terrain be aware of the hypothermia risk. With increasing elevation, temperature decreases and wind chill increases. Avoid steep areas prone to rockfalls and avalanches. Waterfalls are often slippery and hazardous; stream beds have the danger of flash floods.

When working on mountainsides or remote areas, it is particularly important to inform someone of your intended route. Any work on mountains should comply with the usual rules of mountain safety and the normal items for safe mountain walking should be taken.

Landmines continue to pose a severe danger in several countries in Africa, particularly Angola, Zimbabwe, and those in the Horn of Africa, because of decades of conflict and instability (ICBL 2024). In doing fieldwork in these countries, it is incumbent on fieldwork participants to avoid interfering in any way with suspicious metal objects. The presence of any suspicious objects should immediately be reported to the authorities.

During visits to nuclear power plants the first request should be: "a rundown of the safety instructions". These instructions should be taken very seriously. The main safety concern is the emission of uncontrolled radiation into the environment which could cause harm to humans both at the reactor site and off-site.

Landfills and transfer stations can be particularly hazardous both for onsite workers and visiting or temporary personnel such as field researchers who may be unacquainted with the site. "Landfill management" usually helps visitors understand the hazards they may encounter and get them *"au fait"* with safety and emergency procedures.

As telecommunication networks may not be extensive in remote areas the use of satellite-based devices are suitable alternatives to deploy in a multi-layered approach to communication.

Vehicles and machinery should be well maintained and regularly checked (e.g. pre-start) and confirmed to be in sound condition before departing for fieldwork. Ensure that the vehicle or machinery is appropriate for the kind of terrain and the task ahead. The typical desert heat in North Africa may sometimes be almost unbearable and thorough preparation is necessary. Always wear a hat and make sure you carry enough water and food with you for the duration of your trip plus additional water and food as safety precautions.

Useful ways of reducing (but never, of course, totally eliminating) the hazard in difficult environments should be included in the fieldwork plan. It is important to have an emergency plan in place should something go wrong, be it a fire, injured worker or a lost site visitor.

Sampling in Coastal Locations

The African coastline presents special problems in relation to research activities conducted along it; and attention should be paid to whatever safety procedures and regulations apply to such activities. Fieldwork participants should make sure they do not get cut off by the tides, since trapping by tides is always a real possibility (See: Davies 2018). Tides have to be checked when going to critical areas and every effort made to work during low tide when shore ledges are best exposed for safe study. In case of being caught on a rising tide, ensure that you have a means of retreat. Do not go down to low ledges near the sea in stormy weather.

Tidal charts should be consulted before undertaking any shoreline work. Headlands and promontories should be identified, since these may cut off an exit with incoming tides. Attention should be given to the occurrence of riptides and undertow.

Sampling Areas of New and Sudden Disease Outbreaks

The term *outbreak* is used to describe a sudden increase in occurrences of a disease when cases are above what is normally expected for the location or season. An outbreak may be confined to a small and localized group or may affect thousands of people across a continent. The number of cases varies according to causal factor(s), and the size and type of previous and existing exposures.

Epidemiologists and public health professionals tasked with investigating outbreaks of unknown aetiology face several challenges as noted by Goodman et al. (2012). Little is known about outbreaks that occur in remote settings (e.g. in rural areas of Africa) other than what is reported in the global scientific literature. According to the experience of Perrocheau et al. (2023), and a thorough review of the global literature up to 2023, little information is available to the scientific community about outbreaks where the cause is unknown or where there is controversy around the primary agent involved. Again, too, we still do not have detailed information regarding the investigation processes in such outbreaks. Investigation teams face several challenges in data collection, sample collection, storage and shipment (Perrocheau et al. 2023). Davies (2024) has proposed the inclusion of Medical Geologists in investigation teams in order to allay some of these challenges.

As of today (2025) there is still little documentation on how to propose control measures while there is still uncertainty about the cause of the outbreak. In order to assist countries in addressing the issue of outbreaks of unknown aetiology in remote and complex settings, Perrocheau et al., as recently as 2023, designed the "WHO Outbreak Toolkit Project" which sought to develop guidance and tools to improve the quality and timeliness for initial data collection and investigation of outbreaks of unknown aetiology. The toolkit comprises several tools, such as: A WHO manual for investigating outbreaks of possible chemical aetiology (WHO 2021a, b), guidance for investigating clusters of respiratory disease of unknown aetiology, a number of other syndromes (WHO 2018) and a questionnaire for early investigation of outbreak of unknown aetiology (Perrocheau et al. 2021), among others. Davies (2024) has put forward cogent arguments why the inclusion of Medical Geologists in teams investigating outbreaks of diseases of unknown aetiology (DUA) would be invaluable in identifying diagnostic indicators.

Safe Handling of Human and Animal Tissues

In investigating certain Medical Geology phenomena, the practitioner may become involved in sampling and analyses of human (and other animal) tissues. Such work exposes the researcher to many risks, such as to zoonotic pathogens (See, e.g., UCL 2024). For example, in researching the much-misunderstood phenomenon of *geophagy* (or *geophagia*), there may arise the necessity of collecting and analysing stool and blood samples for parasitological examination and haemoglobin iron level. These determinations make possible an evaluation of the role of geophagy in the transmission of geohelminth infections and interference of iron absorption, respectively. Bioassay sampling to determine the uptake of radioactive material by researchers doing radiological sampling also involves collection of urine, faeces, and blood samples for analysis.

It is now clear, certain metals (e.g., copper and zinc) play key roles in cellular function, although several other metals found in the tissues have no known function and are even toxic when present in excess. Toxic metals are absorbed, mainly in the diet, and are unavoidable contaminants of the environment. Determining the content of these metals (whether they are of nutritional value or are toxic) to the human body, as well as the extent to which they are bioavailable, may require tissue sampling and analyses.

It is therefore imperative for the Medical Geologist working with human biological material to be *au fait* with the cultural, ethical, legal and, quintessentially, health and safety considerations when handling human (and other) animal tissues; and abide by established policy recommendations (See, e.g., US CDC 2024) for performing this exercise. Pre-screening of samples with laboratory testing can confirm the presence or absence of specified pathogens.

Before undertaking any work with human or animal tissue, it is necessary to make sure that the nature of the work conforms with the appropriate medical-ethical and animal-experiment legislation and guidelines. It may be necessary to seek approval from the relevant regulatory authorities and/or individuals. Needless to say that investigations involving the handling of human and animal tissues should be guided by an appropriate specialist such as a microbiologist or a biochemist, working in concert with the Medical Geologist (See section on: "Working with Interdisciplinary Colleagues", this Chapter).

It is good practice to treat all human and animal biological samples as potentially infectious, unless they have been inactivated. All work with human and animal biological samples should have an authorised risk assessment in place before work starts. The goal of a safety programme is to lower the risk to as close as possible to zero, although zero risk is as yet unattainable as long as patient specimens and live organisms are manipulated. The reader is referred to the manual of the US Centres for Disease Control and Prevention (US CDC 2024) which provides a comprehensive review of health and safety precautions to be followed in handling of human and animal tissues.

Wearing of PPE is an absolute essential. Lab coats should be worn to avoid soiling the researcher's clothes and to prevent them from acting as a vehicle of infection. Cuts and broken skin must be covered with waterproof dressings.

Disposable unpowdered nitrile gloves should be worn at all times when handling human material in the laboratory. This is particularly important when handling high risk samples. Should the hands become contaminated, they must be washed immediately and surely before leaving the laboratory or moving to a desk or computer within the laboratory.

While the biosafety recommendations are mainly intended to provide maximal protection of human health of the researchers and the environment, many of the proposed health and safety measures will also directly benefit the quality of research activities involving human and animal tissues.

Conclusions

Medical Geology fieldwork is unique, because it combines the methodologies and field techniques of several allied disciplines to reach valid conclusions; an approach which at the same time, encourages collaboration and serves as a platform for innovation.

The issues addressed in this Chapter are significant for all Medical Geologists about to engage in fieldwork in Africa, especially given that while many books that deal with fieldwork or undertaking field-based studies in the geosciences do so in a much broader context focussing on geoscience field methodologies and techniques; and making only cursory references to health and safety issues. This Chapter therefore has sought to address some of the major gaps in the successful conduct of Medical Geology field research and study, especially in Africa, where more severe geographical challenges, in addition to health and safety issues in field research abound, in comparison to most other regions of the world.

The uniqueness and multi-disciplinary experiences encountered in Medical Geology fieldwork brought out in this discourse, will, hopefully, provide the groundwork for graduate students and allied researchers preparing to embark on this very exciting field endeavour!

References

Barnekow U, Fesenko S, Kashparov V, Kis-Benedek G, Matisoff G, Onda Y, Sanzharova N et al (2019) Guidelines on soil and vegetation sampling for radiological monitoring. Issue 486 of the Technical Reports Series, International Atomic Energy Agency (IAEA). ISBN: 9201022182, 9789201022189. 247 pages. https://www-pub.iaea.org/MTCD/Publications/PDF/DOC_010_486_web.pdf. Accessed 02 Oct 2024

Davies TC (2018) A code for geoscientific fieldwork in Africa: guidelines on health and safety issues in mapping, mineral exploration, geoecological research and geotourism. Nova Science Publishers, New York, 325 pp. ISBN: 978-1-53613-033-1. https://www.amazon.com/Code-Geo scientific-Fieldwork-Africa-Geoecological/dp/1536130338. Accessed 04 Oct 2024

Davies TC (2024) Medical geology of Africa: a research primer. Elsevier. ISBN: 9780128187487. https://shop.elsevier.com/books/medical-geology-of-africa/davies/978-0-12-818748-7. Accessed 09 Oct 2024

Goodman RA, Posid JM, Popovic T (2012) Investigations of selected historically important syndromic outbreaks: impact and lessons learned for public health preparedness and response. Am J Public Health 102(6):1079–1090. https://doi.org/10.2105/AJPH.2011.300426. Accessed 25 Aug 2025

ICBL (International Campaign to Ban Landmines) (2024) Landmine Monitor 2024. ICBL-CMC, Geneva. https://backend.icblcmc.org/assets/reports/Landmine-Monitors/LMM 2024/Downloads/Landmine-Monitor-2024-Final-Web.pdf. Accessed 25 Aug 2025

ISO (The International Organization for Standardization) (2017, Updated 2022) (ISO) standard ISO 18400-103:2017, Soil Quality: sampling, Part 103: Safety, 32 pp. https://www.iso.org/standard/62363.html. Accessed 08 Oct 2024

Levain R (2023) Embracing the power of the multidisciplinary approach breaking boundaries and fostering innovation. Short Communication. JBR J Interdiscip Med Dental Sci 6(3):39–42. https://www.openaccessjournals.com/articles/embracing-the-power-of-the-multid isciplinary-approach-breaking-boundaries-and-fostering-innovation.pdf. Accessed 06 Oct 2024

Perrocheau A, Brindle H, Roberts C, Murthy S, Shetty S, Martin AIC, Marks M, Schenkel K, "Minimum Variables for Outbreak Investigation Working Group of the WHO Outbreak Toolkit project" (2021) Data collection for outbreak investigations: process for defining a minimal data set using a Delphi approach. BMC Public Health 21(1):2269. https://doi.org/10.1186/s12889-021-12206-5. Accessed 0 Oct 2024

Perrocheau A, Jephcott F, Asgari-Jirhanden N, Greig J, Peyraud N, Tempowski J (2023) Investigating outbreaks of initially unknown aetiology in complex settings: findings and recommendations from 10 case studies. Int Health 15(5):537–546. https://doi.org/10.1093/inthealth/ihac088. Accessed 0 Oct 2024

Statista (2024) Share of rural population in Africa from 2000 to 2023. https://www.statista.com/statistics/1491165/africa-share-of-rural-population/. Accessed 24 Aug 2025

UCL (The University College of London) (2024) Working with human and animal derived samples. https://www.ucl.ac.uk/safety-services/policies/2023/jun/working-human-and-animal-derived-samples. Accessed 11 Oct 2024

University of Bath (2025) Fieldwork safety standard. https://www.bath.ac.uk/legal-information/fie ldwork-safety-standard/. Accessed 24 Aug 2025

UoS (The University of Stellenbosch) (n.d.) Determining the level of risk (LOW RISK). https://www.eng.sun.ac.za/wp-content/uploads/How-do-I-know-if-my-application-will-be-classified-as-low-medium-or-high-risk.pdf. Accessed 07 Oct 2024

UoS (The University of Stellenbosch) (2024) Risk management: principles and practices. https://shortcourses.sun.ac.za/courses/c-9/2022-5511.html. Accessed 10 Oct 2024

US CDC (United States Centres for Disease Control and Prevention) (2024) Biosafety in microbiological and biomedical laboratories (BMBL), 6th edn. CDC Laboratories. Accessed 05 Oct 2024

US EPA (United States Environmental Protection Agency) (2012) Sample collection procedures for radiochemical analytes in environmental matrices. Office of Research and Development, National Homeland Security Research Center. https://19january2017snapshot.epa.gov/sites/production/files/2015-07/documents/sample_collection_procedures_for_radiochemical_anal ytes.pdf. Accessed 02 Oct 2024

WHO (World Health Organisation) (2018) Protocol to investigate non-seasonal influenza and other emerging acute respiratory diseases, Geneva, World Health Organization. Protocol to investigate non-seasonal influenza and other emerging acute respiratory diseases (who.int). Accessed 0 Oct 2024

WHO Outbreak Toolkit Project (2021a) Data collection for outbreak investigations: process for defining a minimal data set using a Delphi approach. BMC Public Health 21(1):2269. https://doi.org/10.1186/s12889-021-12206-5. Accessed 0 Oct 2024

WHO (World Health Organisation) (2021b) Manual for investigating suspected outbreaks of illnesses of possible chemical etiology: guidance for investigation and control. World Health Organization, Geneva. Manual for investigating suspected outbreaks of illnesses of possible chemical etiology: guidance for investigation and control (who.int). Accessed 07 Oct 2024

Chapter 8
Research Gaps in Knowledge and Future Perspectives on Medical Geology of Africa

T. C. Davies

Abstract The field of Medical Geology in Africa has witnessed a phenomenal growth during the last decade because of the increasing awareness of the role of geoenvironmental variables as causal cofactors in the incidence of some common environmental diseases, including those of unknown aetiology (DUA). This relationship has triggered a surge of research projects on the Continent during the last decade, especially on aspects regarding decipherment of the nature of geochemical circulation of both nutritional and potentially toxic elements (PTEs) in the groundwater-soil-food crop continuum; and how this distribution shapes the nutritional quality of the African diet, and hence maintenance of the proper balance of micronutrients (and vitamins) in metabolic processes. A lot of success has been documented during the last decade in explaining these relationships; but admittedly, more research needs to be done. This chapter presents a cursory summary that encapsulates the book's main features, focussing on gaps in knowledge that remain to be tackled and how this could be effectively done. Finally, a look is taken at future perspectives and prospects.

Introduction

In Africa, environmental pollution, including soil, water and air pollution, contributes to the burden of diseases, a significant proportion of which (diseases) can be linked to geoenvironmental factors. These diseases include respiratory illnesses, water-borne and vector-borne diseases, diarrhoeal diseases, as well as cardiovascular and neurological diseases. *Despite the substantial body of work that has been carried out during the last decade on various aspects of the environmental pollution of Africa's surface environment (soil, water and air) and the resulting maladies, there are several remaining knowledge gaps.*

T. C. Davies (✉)
Faculty of Applied and Health Sciences, Mangosuthu University of Technology, Umlazi, KwaZulu Natal Province, Republic of South Africa
e-mail: daviestheophilus2025@yahoo.com

© The Author(s), under exclusive license to Springer Nature Switzerland AG 2026
T. C. Davies (ed.), *Recent Advances in Medical Geology Research in Africa: A Decadal View*, SpringerBriefs in Earth System Sciences,
https://doi.org/10.1007/978-3-032-13754-8_8

A major objective of most pollution evaluation surveys is to provide a basis for determining the extent of pollution at a specific locality ("hotspot"). For such surveys to be effective, the establishment of reliable environmental quality criteria for soils, sediments and ground- and surface waters is essential to provide a basis for determining the extent of pollution at such localities where we have elevated concentrations relative to "ambient conditions". The term "ambient conditions" (synonymous with "local background") is defined as concentrations of pollutants around a site, but which are unaffected by site-related activities. The "geochemical background" should always be taken into account when determining environmental quality criteria; *however, the terminology relating to "geochemical background" in Medical Geochemistry is yet to be systematised; and reliable and plausible methodology for establishing geochemical background concentrations still need to be worked out.*

In Africa, studies of soil, water and air pollution have been carried out to determine the source of the hazardous substances—primarily from anthropogenic activities like agriculture, mining, industrial processes and leachates from dumpsites; but also, from natural sources such as from volcanic eruptions. Proper source identification and attribution are important and should constitute the first steps in the bid to formulate effective pollution abatement measures.

Addressing issues of environmental pollution in Africa is vital for improving public health and promoting sustainable development. This requires addressing the root causes (sources, mechanisms of transport and sinks) of pollution, improving sanitation and implementing policies to mitigate the negative impacts of environmental degradation (Awewomum et al. 2024). The prospect for addressing these issues is great, as global interest in environmental pollution research during the last decades witnessed a surge, and, thankfully, a favourable shift in international project funding policies.

Soil Pollution

The sources of soil pollution in Africa are multifarious, and include mining, roadside emissions, agricultural activities, auto-mechanic workshop effluents, leachates from refuse dumps and e-waste (Fayiga et al. 2018). In some large oil-producing African countries such as Nigeria and Angola, oil spills constitute a significant source of soil pollution. *Heavy metal exposure, particularly from mining, can lead to various health issues, including reproductive problems, renal dysfunction and neurological damage. A huge research effort supported by international funding has been mounted to address this, through measures such as phytoremediation, but much more needs to be done.* Demarest and Langer (2018) pointed out the existence of important knowledge gaps in soil pollution research, with regard to the deployment of landmines in Africa; and attributed this partly to the unavailability of requisite scientific data because of armed conflicts. *Landmines have been used widely in some of these conflicts, but their impact has not yet been fully investigated as either point or diffuse sources of soil pollution.*

Water Pollution

In Africa, various human activities contribute to water pollution, including agriculture, mining and urbanisation. Substantial advancements are being made in identifying and removing harmful pollutants using innovative technologies and approaches; and research on water chemistry and pollution abatement during the last decade has flourished. This has resulted from the need to understand health risks posed by unacceptably high levels of toxic elements in potable water supplies, as well as the realisation by all stakeholders that climate change would exacerbate the already dire water quality situation that currently exists on the Continent.

A key knowledge gap in Africa regarding water pollution is the insufficient spatial and temporal distribution of monitoring networks for water resources. Consequently, there is a lack of comprehensive data on water quality and pollution levels across several regions and time periods. There is also the unavailability of standardised protocols in many cases for monitoring, quality control, data analysis and decision-making related to water pollution.

To combat water pollution in Africa, according to Bruce and Ma (2021), the implementation of a set of pollution inventory methods is required to enable the assessment of waste loads from pollution sources using functional variables and pollution intensities. The subject of acid mine drainage (AMD) continues to be of immense ecological concern, particularly in South Africa, despite being one of few African countries with established institutional structures for articulating waste disposal policies. *Measured success has been recorded in addressing environmental issues regarding AMDbut several knowledge gaps exist. The extent, impacts, and treatment technologies for AMD are not yet fully understood, which hinder comprehensive solutions and sustainable development goals* (Masindi et al. 2022; Baloyi et al. 2023, 2024; UNEP 2024).

Of the numerous methods that have been developed and researched for combatting AMD, few have been successfully implemented and proven on a large scale over extended periods. *These knowledge gaps highlight the need for more research and development centred around long-term monitoring, the potential for by-product recovery to offset treatment costs and full-scale implementation of new technologies.*

Air Pollution

Despite the increasing awareness that has produced a wave of investigative studies on air quality, there remains a dearth of air quality data in Africa. Burning fossil fuels, industrial emissions and household air pollution from solid fuels (like charcoal and kerosene) are major contributors to air pollution. Such pollution is linked to respiratory illnesses, cardiovascular diseases and even lung cancer.

The burst of scientific activity on air quality in Africa during the last decade is in accord with global activities in the field of atmospheric chemistry, to address climate change and health issues. The air quality situation is most serious In the Sub-Saharan Africa region where most of the Continent's mining activities are concentrated, especially with regard to the somewhat underrated problem of radon entry into homes, and its accumulation in deep underground mines.

Gaps in knowledge about air pollution in Africa also abound in the area of data collection and regulation. Many African countries lack comprehensive air quality monitoring systems, especially in densely populated urban areas (see, e.g., Katoto et al. 2019; Kalisa et al. 2023). This deficiency hinders accurate assessment of pollution levels, understanding its health impacts and implementing effective mitigation strategies.

A 2019 UNICEF Report, appositely summarises the present situation regarding air quality, noting that "… only about 6% of children [in Africa] live near reliable, ground-level monitoring stations that provide real-time data on the quality of air they are breathing …". In Europe and North America, on the other hand, about 72% of children live near reliable monitoring stations (UNICEF 2019). There is therefore a great need to increase the base of reliable, local, ground-level measurements for effectively responding to this little understood wanton killer of children across the Continent.

Radon Exposure

Exposure to airborne radon progeny not only in the domestic environment but also in deep underground mine settings yields the largest source of exposure to ionising radiation of the population. The greater awareness of the severity of the health consequences of radon exposure in homes and deep underground mines of Africa has brought about a rise in the number of studies on "Radon in Africa" in the last decade (e.g., Botha et al. 2018; Le Roux et al. 2019; Radebe and Mathuthu 2023; Nunes and Curado 2025). Most of these studies address radon exposure situations and health effects in South Africa.

Despite the surge in research on radon exposure in Africa, particularly near gold mine tailings in South Africa, significant knowledge gaps remain. The dearth of studies on radon concentrations and its exposure in Africa can be attributed to a lack of awareness among policymakers and the public, insufficient expertise in radiation protection and measurements and restricted access to resources such as laboratories and testing equipment. *A lot more research needs to be done to tackle radon exposure problems in African countries, especially in homes located in the neighbourhood of uranium and gold mining centres and over large areas of felsic intrusive rock.* A comprehensive understanding of radon's impact on public health and the effectiveness of mitigation strategies are issues that need to be further researched (Nunes and Curado 2025).

Criticality of the Intake Level for Geomedically Important Elements

An important issue often neglected by many food safety and food quality regulators is the full awareness of the range of natural variability, viz., that not only high values of geogenic toxicants and anthropogenic pollution may represent a threat to human health. On a regional scale, element deficiency-related health problems may well be the more important human-health problem, as was discussed in Chap. 3 of this book. Studies at the lower end of background variation and setting lower, minimum admissible concentration values might have a greater impact on the overall health of humans than the present focus on high values and toxicity. Documenting and understanding natural variability are therefore of high importance in Medical Geology as well as in other environmental studies.

It is hard to recognise and understand changes in natural systems if we do not understand the range of baseline levels. Systematic geochemical mapping is apparently the best method of providing information about natural variability of geochemical background (see under: "Introduction", this chapter). In this connection, geochemical maps provide a valuable tool in addressing a wide range of environmental problems at the local to national scale, and for identifying potential areas of element deficiency or excess.

Examples abound of the wealth of valuable information generated during several decades of geochemical mapping by national geological surveys and related organisations throughout Africa. However, many of these datasets cannot readily be applied to broader regional or global studies because there was no standardisation of the sampling and analytical protocols used. Concept of "geochemical background" (defined earlier under "Introduction", this chapter) is therefore of high importance in our studies on sufficiency or excess in terms of element intake levels, even of the "essential elements".

Bioavailability refers to the fraction of a substance that is absorbed into the bloodstream and reaches its target site of action in the body, whereas *bioaccessibility* refers to the fraction of a substance that is released from its matrix (such as food) and is readily available for absorption in the gastrointestinal tract. These definitions are quite exquisite, for they embrace the concept of multi-functionality, and can apply to contaminants being available or accessible to a range of organisms, including higher animals, by simply addressing supply across the membrane of the organism in question.

Bioavailability and bioaccessibility are complex issues that need to be assessed in a circumspect manner in order to determine whether or not there may be adverse effects for human health (Petruzzelli et al. (2020); meaning that cognisance has to be taken of the various factors such as soil properties, interactions with other substances and the specific conditions of the site, rather than relying only on simplified laboratory tests (Peijnenburg and Jager 2003). A holistic approach to their measurement should therefore be adopted, with the results being put into the context of the whole geochemistry. Although there is still a great variability in the determination of bioavailable and bioaccessible fractions, an assessment of the real risks for humans and the environment must be based on these parameters and not on the total element concentration.

A very important limitation on our ability to estimate bioavailability is the scarcity of data that provide a direct estimate of the human absorption of PHEs. This limitation can be overcome in part using the most representative animal models. Physical and chemical parameters need to be incorporated in predictive models which relate toxic metals in soils to human health.

It is important to note that exposure, dose and effects are not independent parameters but rather successive phases of a continuum linking the soil to human health through bioavailability and bioaccessibility processes. *However, there is yet very limited documented research on the speciation of PHEs in the simulated gastrointestinal systems.* Understanding the mechanisms by which simulated gastrointestinal solutions alter the species present could enable the performance of a proper human health risk assessment, a topic that will need to be researched with much more rigor in future investigations.

Artisanal and Small-Scale Mining in Africa

Artisanal and small-scale mining (ASM) is an important sector in African countries with a mining history. As much as it can provide a vital source of income and livelihood for many, especially in rural areas, it also poses environmental and health risks due to mercury use, other toxic metal releases and unregulated practices. Given its well-established impacts on human health, its inherent toxicity, as well as its tendency to accumulate in living organisms, mercury pollution is of grave concern to humanity. Notable known toxicity due to inhalation of the vapour of this non-essential element includes neurological and carcinogenic effects, respiratory issues and kidney problems.

Several studies have focussed on the integration of ASM into the formal economy, and on provision of support and training to ASM miners, which, admittedly, are crucial steps towards sustainable and responsible mining. *However, research on promoting mercury-free technologies and the element's elimination in humans, which are of considerable importance, have been given relatively little attention; even though results from such research can be used to improve management of human health and environmental risks from mercury exposure.*

Measuring progress and the impact of support interventions on the development of the African ASM sector is no mean task, due largely to the dearth of data on the sector and the interventions themselves. *In the few available studies, there is the lack of reliable data with respect to the exact number of people employed by the sector, range of activities, geographical distribution, demographic profiles and the level of contribution to the economy. There is the need for research to assess the environmental health impact of past and existing interventions on the ASM sector to draw lessons for future development.*

The Fate of Inhaled Geogenic Dust Particles in the African Environment

In the African environment inhaled dust particles can have significant health effects, particularly respiratory disorders. The fate of inhaled particles depends on their size, with smaller particles ($PM_{2.5}$) penetrating much deeper into the lungs where they may cause inflammation, oxidative stress and tissue damage. Larger particles are typically trapped by the respiratory system's natural defences, such as mucus and cilia, and expelled. There are other factors that can influence the potential respiratory toxicity of inhaled particles, including their mineralogy, chemical composition, morphology and surface characteristics and the capacity to generate harmful free radicals (Tomašek et al. 2025).

Inhalation of asbestos dust in South Africa is a serious public health issue due to the Country's history of asbestos mining and processing. *Despite years of research on asbestos exposure and related diseases in Africa, significant knowledge gaps still exist particularly in understanding the extent of occupational exposure, the prevalence of asbestos-related diseases such as asbestosis, lung cancer and mesothelioma, and the effectiveness of current regulations and interventions.* This situation remains extant, despite the continued use of asbestos in various sectors, including mining, manufacturing and construction. Investments in research are needed to generate reliable data on the prevalence of asbestos-related diseases, occupational exposure and the effectiveness of interventions.

The Saharan Dust Plume (SDP) comprises large clouds of dust that travel thousands of kilometres to the Atlantic Ocean or Mediterranean Seas. The SDP contains nutrients and minerals, including iron, necessary for photosynthesis and other cellular functions. The Plume is known to degrade air quality, posing serious health threats to humans, especially those with pre-existing lung conditions. These issues include respiratory and cardiovascular diseases and even death in extreme cases.

Despite this knowledge, large uncertainties exist in our ability to predict future trends in Sahara dust emissions and model-projected atmospheric circulation patterns, a lack of detailed information on the chemical composition of dust particles, and insufficient understanding of the long-term health effects, in dust-affected regions.

Research on Dust Control Mechanisms

Most of the research on dust control in Africa, particularly South Africa, has been on dust emission from mining and exposure of miners and associated communities to prevent lung diseases such as silicosis and coal workers' pneumoconiosis (CWP). This research has focussed on the development and implementation of effective dust control systems, improvement of monitoring techniques and the enhancement of worker training and awareness. On 16 May 2025, the Minister of Forestry, Fisheries and the Environment (MFFE) of South Africa published revised regulations aimed at further improving dust control measures, *inter alia* (Baille et al. 2025).

In Africa, current gaps in knowledge exist on approaches for successfully controlling dust emissions and mitigating human exposure, especially in dynamic environments such as mining centres and regions prone to sand and dust storms. The shortage of information regarding Sahara dust transport and its health consequences stem from the complexity of the interactions between Sahara dust particles and respiratory health (Georgakopoulou et al. 2024). *This underscores the existence of critical research gaps that need attention to better understand and manage the health risks related to such dust.*

Additional studies are also needed in the improvement of real-time monitoring structures, targeted countermeasures for Sahara dust, and an enhanced understanding of the health impacts of dust exposure in general. Furthermore, the specific impacts of Saharan dust on different ecosystems and the interactions between the dust and local meteorological situations remain areas of active research.

Minimising Dust Exposure

To minimise inhaled dust exposure in mining, a multifaceted approach is crucial. Such an approach combines engineering controls, administrative controls and the use of personal protective equipment (PPE), which include respirators and masks. Using a mask reduces the number of particles that might be inhaled, since a significant proportion of mineral particles are usually of relatively large dimensions. However, personal protective equipment should not be a substitute for proper dust control and should be used only where dust control methods are not yet effective or are inadequate.

Staying indoors as much as possible is a viable option, especially during periods of high dust concentration. Moreover, keeping windows and doors closed will drastically minimise the amount of dust from entering your property.

The Health Impact of Climate Accentuated Geological Processes in the African Coastal Zone

Climate-induced perturbations in geological processes, particularly sea-level rise, pose significant threats to Africa's coastal zones, exacerbate the spread of diseases and drastically affect food security and food safety. Increased sea levels lead to salt-water intrusion and storm surges that can damage infrastructure, displace populations, and in the long run contaminate freshwater sources and create favourable breeding grounds for disease vectors like mosquitoes. This, combined with rising temperatures, can expand the geographic range and transmission seasons of vector-borne diseases.

However, despite this knowledge, limitations exist in our ability to forecast the climate and disease, including how we incorporate changes in human behaviour and how we attribute climate/weather events solely to an infectious disease outcome.

Food systems, which encompass agricultural production, processing, distribution and consumption, are significantly impacted by both natural and human-caused climate change. The increased rates and frequencies of geological processes under a climate change scenario can exacerbate existing challenges within food systems, creating a complex interplay between geological processes, climate and food production. *The uncertainty around the impacts of climate change needs urgent attention, and more interdisciplinary research is vital for deciphering the unknowns.*

Health and Safety Issues in Medical Geology Fieldwork

All stakeholders concerned with Medical Geology fieldwork in Africa should pay cautious attention to health and safety to prevent injuries and illnesses (Davies 2018). Fieldworkers should adopt a risk-based approach to the management of health and safety risks and set out reasonably practicable actions to avoid them (Tshwane: Higher Health 2021). All fieldwork carried out should be subject to a Risk Assessment and Approvals Procedure (RAAP). Such a procedure involves a stepwise process, from fieldwork planning, through hazard identification, risk profiling, risk assessment and approvals, followed by monitoring and review.

Key practices include staying hydrated, wearing appropriate protective gear like safety glasses and hard hats, being aware of environmental hazards such as unstable cliffs, and informing someone of your route and schedule. Maintaining a healthy lifestyle, including proper nutrition and rest, is also crucial for optimal performance and well-being during fieldwork.

Fieldwork must be properly managed so that the institution can demonstrate that it has done all that is reasonably practicable to minimise risks to health and safety; and to reduce the likelihood and ameliorate the consequences of any reasonably foreseeable accident during fieldwork. It is the duty of university establishments and research institutions to provide structured support systems for interpersonal skills training of fieldworkers, and create avenues for discussion, reflection and experience sharing to enable fieldworkers to navigate the practical and ethical challenges they face (Kombe et al. 2019).

Future Outlook

Despite the substantial advances that have been made in Medical Geology during the past decades, much more needs to be done to maintain the momentum. The important short-term goals include:

- Increased interaction and communication between the geoscience, biomedical and public health communities to protect human health from the toxic effects of physical, chemical and biological agents in the environment.
- A gratifying trend is the increasing recognition and acceptance of Medical Geology and its benefits by politicians, decision-makers and the public. Such trend is set to continue, for this is vital for the continued success of this emerging science. Such recognition is achieved, *inter alia* through various modes of communication, such as by the wide distribution of *policy briefs* to government policy-makers, municipal authorities, relevant NGOs and other stakeholders including the public, emphasising the value of producing complete geochemical maps of the various regions to foster recognition and acceptance.
- Degree programmes—Encourage widespread introduction of Medical Geology curricula in environmental science education at tertiary level institutions. Indeed, the past decade has witnessed the incorporation of Medical Geology modules by an increasing number of African geoscience and public health institutions in their undergraduate curricula, *in tandem* with the appearance during this period of highly customised textbooks and primers (e.g., Davies 2019, 2024; Prasad and Vithanage 2023). These texts have united medical, geological and environmental insights in one continuous approach to public health and have provided an essential overview of the field for advanced students as well as medical and environmental researchers wishing to explore the complex relationships between the geological environment and human health. Many students retain their fascination and interest after their undergraduate exposure to the subject by undertaking postgraduate research and study, thereafter. Proposed graduate programmes and customised curricula have already been designed and published (see, e.g., Davies 2019), auguring the appearance of more M.Sc. and Ph.D. degree courses. Certificate programmes at universities and colleges around Africa are also urgently needed.
- The provision of generous sources of funding for researchers and internships and pre-and post-doctoral training opportunities would ensure continuity of the upward trajectory of Medical Geology research in Africa.
- Common multidisciplinary outlets for journal articles and conferences should be sought. To date only a handful of specialised journals (e.g., "Environmental Geochemistry and Health" and "GeoHealth" cater specifically for publications in Medical Geology.

- Success stories—It is imperative that African Medical Geologists clearly demonstrate the value of their input to the biomedical and public health communities. Successes attained in credible aetiological attribution by mapping disease distribution overlain with geoenvironmental correlates should be emphasised.
- Jobs: The primary applications of Medical Geology are in research, public health and environmental pollution. Medical Geology graduates of Africa will continue to provide the requisite skills for employment in professional geoenvironmental and public health consultancies and government environmental agencies. Professional Medical Geologists also work in public health consultancies, and health spar resorts. Others become professors or research staff at universities and colleges, engaged in curricula such as Public Health, Toxicology, Environmental Pollution and Forensic Geology. Employers need to be more aware of the benefits of Medical Geology.
- There have been significant improvements in analytical capacity for determination of geomedically important elements (e.g., iodine, fluorine, selenium, zinc, arsenic, lead and mercury) at African geochemical laboratories over the last 10 years, coinciding with the acquisition of smaller, more compact, accurate and easy-to-use analytical instrumentation such as the portable X-ray spectrometer (see: Davies 2023). *However, the need for a critical mass of well-trained analytical geochemists (M.Sc. and Ph.D. holders in analytical chemistry) is still quite apparent.* Such training should engender the ability to (i) install; (ii) operate; (iii) trouble-shoot; and (iv) maintain, modern analytical instrumentation needed for determination of ultra trace amounts of geomedically important elements needed for making tangible conclusions in the field. This is a necessary step that should precede the acquisition of today's analytical instrumentation and other essential infrastructure that should preferably be located in suitable regional centres in Africa. *Although substantial gains realised in analytical capacity within the region have contributed to the increase in self-reliance and sustainability of analytical geochemistry programmes involving ultra-trace elements, many challenges remain and massive investment is required.*

The Africa Geochemical Database Project

The original Project's aim was: "To develop a land base multi-element geochemical baseline database for mineral resource and environmental management" (Darnley et al. (1995). It is submitted that such a dataset would be invaluable for studies in Medical Geology, in addition to having an array of multipurpose, multi-national environmental applications. A high-quality geochemical baseline data will be indispensable in monitoring the geochemical state of the African surface environment, and in identifying problem areas, such as those with nutritional element deficiencies or excess, and high values of PHEs. The Project has suffered considerable setbacks since its inception in the early 1990s. Nevertheless, a couple of successes were recorded in the last few years; but many challenges remain, the major ones of which are listed as follows:

1. Geochemical complexities of the surface environment of Africa make it essential for new standardised sampling and analytical protocols to be developed, re: the Darnley 1995 recommendations. This has not yet been fully achieved.
2. Lack of sufficient analytical capacity (e.g., more modern highly sensitive analytical instrumentation) at regional centres to perform such large-scale ultraprecise undertaking; despite the extant significant level of analytical capacity in some South African universities and research institutions.
3. Lack of sufficient number of highly trained technicians able to install, maintain, trouble shoot and operate today's modern analytical instrumentation.
4. Lack of unanimity among geochemists on determination of "background values".
5. Lack of conviction by African Governments (Geological Surveys), parastatals, other stakeholders and potential donors of the value of geochemical knowledge of Africa's surface geochemical environment; and hence, the lack of investment of large sums of money required to carry out such a large-scale and dedicated exercise.

Conclusion

The future of African Medical Geology is bright. This is clearly evidenced by the voluminous research literature of the last decade and the recent identification of probable "new" correlations between geology and health. Collaboration between geologists and medical scientists is getting stronger, analytical capacity for environmental and geomedical samples at African geochemical laboratories has markedly improved, as is the introduction of graduate Medical Geology programmes in several tertiary institutions all around the Continent. It is possible to suggest that this trend will continue well into the future!

References

Awewomum J, Dzeble F, Takyi YD, Ashie WB, Ettey ENYO, Afua PE et al (2024) Addressing global environmental pollution using environmental control techniques: a focus on environmental policy and preventive environmental management. Discover Environ 2(8). https://doi.org/10.1007/s44274-024-00033-5. Accessed 07 Sep 2025

Baille B, Muller E, Mutula M (2025) South Africa's Ministry of forestry, fisheries and environment invites comments on the draft national dust control amendment regulations. Minister of Forestry, Fisheries and the Environment (MFFE) of South Africa. https://www.hsfkramer.com/notes/africa/2025-.... Accessed 07 Sep 2025

Baloyi J, Ramdhani N, Mbhele R, Ramutshatsha-Makhwedzha D (2023) Recent progress on acid mine drainage technological trends in South Africa: prevention, treatment, and resource recovery. Water 15:3453. https://doi.org/10.3390/w15193453. Accessed 29 May 2025

Baloyi J, Ramdhani N, Mbhele R, Simate GS (2024) Acid mine drainage from gold mining in South Africa: remediation, reuse, and resource recovery. Mine Water Environ 43:418–430. https://doi.org/10.1007/s10230-024-00994-2. Accessed 29 05 2025

Botha R, Labuschagne C, Williams AG, Bosman G, Brunke E-G, Rossouw A, Lindsay R (2018) Radon-222 measurements at cape point: a characterization of a 15-year time series. Clean Air J 28(2):19–20. https://doi.org/10.17159/2410-972x/2018/v28n2a11. Accessed 02 June 2025

Bruce M, Ma (2021) Recent advances on water pollution research in Africa: a critical review. Int J Sci Adv 2(3). https://doi.org/10.51542/ijscia.v2i3.23

Darnley AG, Björklund A, Bølviken B, Gustavsson N, Koval PV, Plant JA, Steenfelt A, Tauchid M, Xie X, Garrett RG, Hall GEM (1995) A Global geochemical database for environmental and resource management: final report of IGCP Project 259. Earth Sci 19:122 (UNESCO Publishing, Paris). https://unesdoc.unesco.org/ark:/48223/pf0000101010. Accessed 26 Aug 2024

Davies TC (2018) A code for geoscientific fieldwork in Africa: guidelines on health and safety issues in mapping, mineral exploration, geoecological research and geotourism. Nova Science Publishers, New York, 325 pp. Amazon.com.au. ISBN: 978-1-53613-033-1. Accessed 28 Apr 2024

Davies TC (2019) A medical geology curriculum for African geoscience institutions. Sci Afr 6:e00131. https://doi.org/10.1016/j.sciaf.2019.e00131. Accessed 30 Nov 2024

Davies TC (2023) Some highlights on applied geochemistry application in Southern Africa during 2022–2023. Report of the Regional Councillor for Southern Africa. Association of Applied Geochemists. EXPLORE 200, pp 18–19. Accessed 06 Dec 2024

Davies TC (2024) Medical geology of Africa: a research primer, 1st edn. Elsevier. ISBN: 9780128187487. https://shop.elsevier.com/books/medical-geology-of-africa-a-research-primer/davies/978-0-12-818748-7. Accessed 28 Apr 2024

Demarest L, Langer A (2018) The study of violence and social unrest in Africa: a comparative analysis of three conflict event datasets. Afr Aff 117:310–325. https://doi.org/10.1093/afraf/ady003. Accessed 29 May 2025

Fayiga AO, Ipinmoroti MO, Chirenje T (2018) Environmental pollution in Africa. Environ Dev Sustain 20:41–73. https://doi.org/10.1007/s10668-016-9894-4. Accessed 22 June 2023

Georgakopoulou VE, Taskou C, Diamanti A, Beka D, Papalexis P, Trakas N, Spandidos DA (2024) Saharan dust and respiratory health: understanding the link between airborne particulate matter and chronic lung diseases (Review). Exp Ther Med 28(6):460. https://doi.org/10.3892/etm.2024.12750. Accessed 22 June 2023

Kalisa E, Clark ML, Ntakirutimana T, Amani M, Volckens J (2023) Exposure to indoor and outdoor air pollution in schools in Africa: current status, knowledge gaps, and a call to action. Heliyon 9(8):e18450. https://doi.org/10.1016/j.heliyon.2023.e18450. Accessed 29 May 2025

Katoto PD, Byamungu LN, Brand A, Mokaya J, Strijdom H, Goswami N, De Boever P, Nawrot TS, Nemery B (2019) Ambient air pollution and health in Sub-Saharan Africa: current evidence, perspectives and a call to action. Environ Res 173:174–188. https://doi.org/10.1016/j.envres.2019.03.029. Accessed11 Jan 2025

Kombe FK, Marsh V, Molyneux S, Kamuya DM, Ikamba D, Kinyanjui SM (2019) Enhancing fieldworkers' performance management support in health research: an exploratory study on the views of field managers and fieldworkers from major research centres in Africa. BMJ Open 9(12):e028453. https://doi.org/10.1136/bmjopen-2018-028453. Accessed 31 May 2025

Le Roux RR, Bezuidenhout J, Smit HAP (2019) The influence of different types of granite on indoor radon concentrations of dwellings in the South African West Coast Peninsula. J Radiat Res Appl Sci 12(1):375–382. https://api.semanticscholar.org/CorpusID:210617301. Accessed 02 June 2025

Masindi V, Foteinis S, Renforth P, Ndiritu J, Maree JP, Tekere M, Chatzisymeon E (2022) Challenges and avenues for acid mine drainage treatment, beneficiation, and valorisation in circular economy: a review. Ecol Eng 183:106740. https://doi.org/10.1016/j.ecoleng.2022.106740. Accessed 29 May 2025

Nunes LJR, Curado A (2025) Indoor radon exposure in Africa: a critical review on the current research stage and knowledge gaps. AIMS Public Health 12(2):329–359. https://doi.org/10.3934/publichealth.2025020

Peijnenburg WJ, Jager T (2003) Monitoring approaches to assess bioaccessibility and bioavailability of metals: matrix issues. Ecotoxicol Environ Saf 56(1):63–77. https://doi.org/10.1016/s0147-6513(03)00051-4. Accessed 17 Apr 2024

Petruzzelli G, Pedron F, Rosellini I (2020) Bioavailability and bioaccessibility in soil: a short Petruzzelli review and a case study. AIMS Environ Sci 208–225. https://doi.org/10.3934/environsci.2020013

Prasad MNV, Vithanage M (eds) (2023) Medical geology: En route to one health

Radebe M, Mathuthu M (2023) An overview of radon emanation measurement system for South African communities. In: Aide M (ed) Rare earth elements—Emerging advances, technology utilization, and resource procurement. https://www.intechopen.com/chapters/85384. Accessed 31 May 2025

Tomašek I, Eychenne J, Damby DE, Hornby AJ, Romanias MN, Moune S et al (2025) Physico-chemical properties and bioreactivity of sub-10 μm geogenic particles: comparison of volcanic ash and desert dust. GeoHealth 9:e2024GH001171. https://doi.org/10.1029/2024GH001171. Accessed 31 May 2025

Tshwane: Higher Health (2021) Planning, preparing for, and conducting fieldwork in the context of COVID-19—A principles-based approach. https://rd.mandela.ac.za/rcd/media/Store/documents/RecH/A-Guideline-for-conducting-Fieldwork-during-COVID.pdf. Accessed 31 May 2025

UNICEF (United Nations Children's Fund) (2019) Silent suffocation in Africa: air pollution is a growing menace, affecting the poorest children the most. https://www.unicef.org/media/55081/file/silent_suffocation_in_africa_air_pollution_2019.pdf. Accessed 02 06 2025

UNEP (United Nations Environment Programme) (2024) Knowledge gaps in relation to the environmental aspects of tailings management. Final Knowledge Gaps Report, UNEP, Nairobi. https://unece.org/sites/default/files/2024-09/Final%20Knowledge%20Gaps%20Report_Environmental%20Aspects%20of%20Tailings%20Management%20%28January%202024%29_1.pdf. Accessed 29 May 2025

Author Index

T. C. Davies (ed.), *Recent Advances in Medical Geology Research
in Africa: A Decadal View*, SpringerBriefs in Earth System Sciences,
https://doi.org/10.1007/978-3-032-13754-8